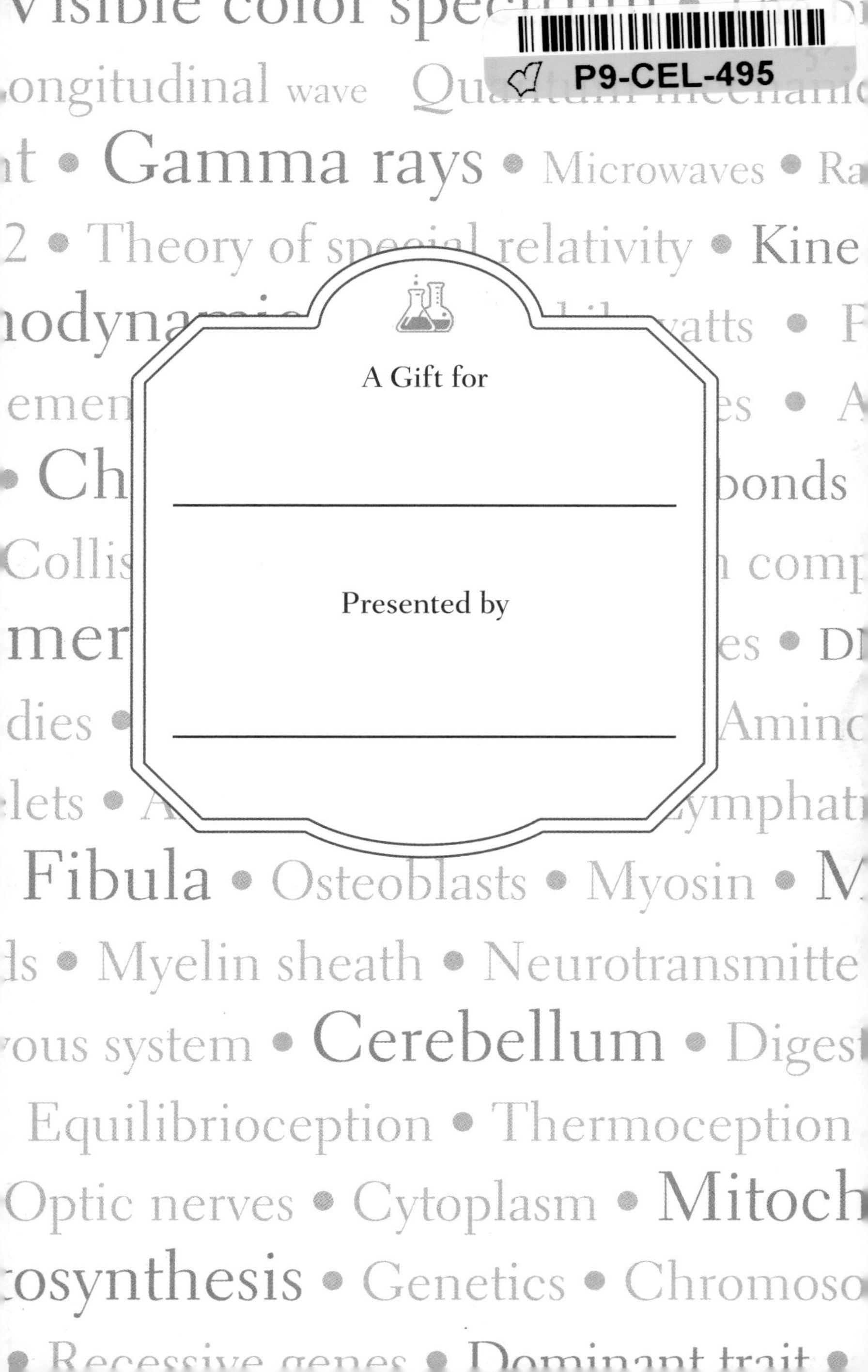
P9-CEL-495
Gamma rays • Microwaves
Theory of relativity
A Gift for
Presented by
Fibula • Osteoblasts • Myosin
Myelin sheath
Cerebellum
Equilibrioception • Thermoception
Optic nerves • Cytoplasm
Genetics
Dominant trait

I Used to Know That

SCIENCE

I Used to Know That SCIENCE

Fascinating Truths About How Animals Evolve, Plants Grow, Brains Work, Molecules Bond, and Stars Explode

Marianne Taylor

Reader's digest

The Reader's Digest Association, Inc.
New York, NY / Montreal

This book is for my family.

A READER'S DIGEST BOOK

First published in Great Britain by Michael O'Mara Books Limited, 9 Lion Yard, Tremadoc Road, London SW47NQ.

FOR READER'S DIGEST
Consulting Editor: Andrea Chesman
Designer: Nick Anderson
Illustrator: Andrew Pinder

ISBN 978-1-62145-279-9

Library of Congress Cataloging in Publication Data
Taylor, Marianne, 1972-
I used to know that : science : Fascinating Truths About How Animals Evolve, Plants Grow, Brains Work, Molecules Bond, and Stars Explode / Marianne Taylor.
p. cm.
"A Reader's Digest Book."
Includes index.
ISBN 978-1-60652-467-1 – ISBN 978-1-60652-469-5 (epub) – ISBN 978-1-60652-468-8 (adobe)
1. Science. I. Title. II. Title: Science.
Q158.5.T39 2012
500–dc23
2011037144

We are committed to both the quality of our products and the service we provide to our customers. We value your comments, so please feel free to contact us.

The Reader's Digest Association, Inc.
Adult Trade Publishing
44 S. Broadway
White Plains, NY 10601

For more Reader's Digest products and information, visit our website:
www.rd.com (in the United States)
www.readersdigest.ca (in Canada)

Printed in China

1 3 5 7 9 10 8 6 4 2

Contents

Introduction

Most of us know science can be fun, and there is an immense sense of achievement when you do finally grasp a tricky concept. But our experiences from school days may keep us from trying, and we switch off our brains the moment things get too technical. School's finished, and there's no need to struggle with this stuff anymore, right?

Well, when you leave school, you can indeed leave behind a lot of the stuff you learned. You can go day after day, week after week without needing to remember the names of the presidents or the world's tallest mountains and longest rivers. You may never need to speak French or German in your life again, and if your command of written English goes downhill somewhat, it probably will go unnoticed by most. Science is different, though. Science is everywhere. It muscles in on your day-to-day life whether you want it to or not. From understanding how your body works to figuring out how best to heat your home, it's all about scientific principles, and you have everything to gain and nothing to lose from familiarizing—or re-familiarizing—yourself with a bit of science.

This book covers more or less the same stuff you'll find in three years of high school science. I'm not trying to get you through an exam; I'm merely hoping to get you excited about

THE SCIENCE TRINITY

The traditional scientific subjects taught in high school are physics, chemistry, and biology.

Physics is the study of fundamental forces and particles. Because it deals with things that we mostly can't see directly, there's a lot of theory and a lot of mathematics. Physics is sometimes considered the study of the very big and the very small, from universes and stars to subatomic particles.

Once you have assembled enough subatomic particles to make an actual atom, you enter the realm of chemistry. This science deals with the properties and behavior of atoms and molecules of the many and varied elements and compounds in various different settings—from the school lab to the blast furnace, from the atmosphere of Earth to the insides of the cells in our bodies.

And so chemistry segues into biology, the study of living things. From the molecular machines that drive our internal processes, biological study works up to the structure and organization of the cell and to the ways cells of different types combine to form the body's various different systems, how those systems work, and finally from the individual organism to the whole ecosystem.

---

science, so you'll find that there may be a bit more about some things and a bit less about others. Some of the concepts we're going to explore are difficult, but every care has been taken to explain stuff in straightforward language and to steer clear of unnecessary terminology, mind-bending mathematics, and overly esoteric subject matter. Wherever possible, diagrams and real-world examples are added to help make things more understandable.

If the sciences didn't work the way they do, we wouldn't be here, and being able to understand how science works is one of the best things about being human.

SCIENCE AND PSEUDOSCIENCE

Because it applies to everything in this book, this seems like the right place to explain what the scientific method is—and isn't. For science to work, it needs to be carried out with an honest and consistent method. The basic process is that a hypothesis (a possible explanation) for a phenomenon is proposed. Then we come up with predictions that would fit that hypothesis, then test the predictions to see if the hypothesis can be proved false.

Pseudoscience covers things that look like science but don't fit the scientific method. For example, we can't scientifically test for the existence of a god, because the statement "God exists" can't be proved false. We can, however, put down any lack of evidence to supernatural explanations: God is omnipotent and can therefore choose to go undetected regardless of our searching methods. Pseudoscience also describes ideas that are claimed to be true but for which convincing evidence has not been found.

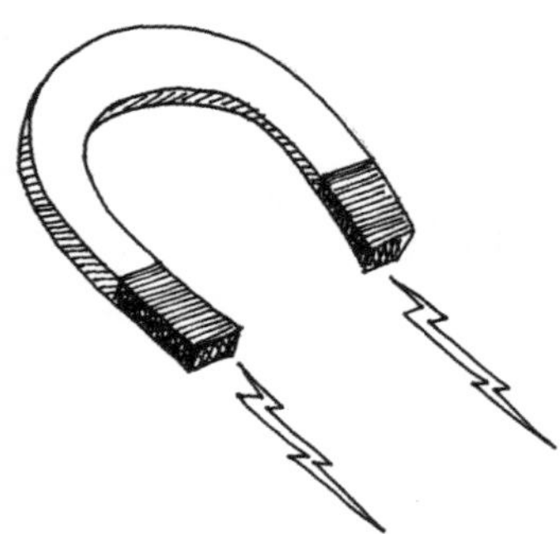

Electromagnetic Force

Strong Nuclear Force

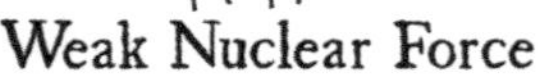

Weak Nuclear Force

Gravitational Force

1

Physics

Four Fundamental Forces

We live in a universe that is subject to a whole host of physical laws, without which neither we nor the world we live in would exist as we know it. Four basic forces act upon matter and determine its behavior. (Remember the definition of matter? It is anything that has mass and occupies volume.) The four forces are the electromagnetic force, the strong nuclear force (or strong force), the weak nuclear force (or weak force), and the force of gravity.

Electromagnetic Force

This force is all about the attraction between dissimilar charges and the repulsion between negative charges. Electricity is the flow of electrons—negatively charged particles. Electrons are a constituent of atoms, as are positively charged protons. The electromagnetic force is what holds atoms together. It also holds together molecules (two or more atoms bonded together; more about this in the Chemistry chapter). Moving charged particles create magnetic fields—so magnetism is related to electricity, and together they form this force. It is the second strongest of the forces and operates over infinite distances.

The Nuclear Forces

The strong and weak forces both act on the particles inside the nucleus of an atom—the positively charged protons and uncharged neutrons. The strong force holds the protons and neutrons together and is the strongest of the four forces. However, its range is small—about the size of an average atom's nucleus. The weak force is the second weakest of the four and operates over a very small range—less than the size of a proton. It acts upon particles even smaller than protons and neutrons and is responsible for nuclear decay—the breaking down of very large nuclei into smaller ones. Without it the nuclear fusion that makes stars burn would not happen. That's a pretty good example of how fundamental these forces are.

Gravitational Force

Gravity is the weakest of the four forces. That may come as a surprise, given that gravity is the one that we are probably most acquainted with, and we know just how firmly it insists that we don't float off into space, but there it is. It does make

up for this feebleness by operating over infinite distances. Gravity is the force that attracts particles or larger bits of matter toward one another.

WEIGHT AND MASS

When we talk about something's weight, we're usually talking about kilograms or grams, pounds or ounces, or tons. These measurements, though, are units of mass rather than weight. Weight is a bit more complex, describing the gravitational forces an object is experiencing due to its mass. The equation for weight looks like this:

weight = mass x *g*

The "*g*" is an expression of the local gravitational pull, which you'd determine by measuring how quickly a free-falling object accelerates. On Earth *g* is about 9.8 meters per second. (It is written as an italic so you don't get mixed up between it and the non-italicized "g," which is short for gram.)

Because we mostly don't take objects off planet Earth in order to weigh them, the terms weight and mass are used interchangeably. However, when we escape the gravitational pull of this planet, the weight of our chosen object changes. In space, far from all other matter, it will essentially be weightless (but not quite—remember that gravity's reach is infinite, so there's nowhere in the universe where there's no gravity at all) because anything multiplied by zero is zero. Take the object to the moon and it has some weight, but less than it does on Earth because the moon is less massive. However, the object's mass doesn't change at all, ever, whether you take it to Berlin or Betelgeuse.

Planets, Stars, and Galaxies

Our universe holds a whole lot of space and concentrated pieces of matter in the form of the various "heavenly bodies." Planets orbit stars to form solar systems, and stars are clustered into galaxies, which usually show an ordered structure and movement. Our galaxy is the Milky Way, and our sun is one of at least 20 billion stars within it. There are some 100 billion other galaxies out there in the observable universe. These are some seriously big numbers we're throwing around. One more biggie before we look at things in more depth: The observable universe is at least 46 billion light-years in any direction. And there is plenty more universe in the unobservable category, too.

Planets

Back to Earth (or thereabouts) for a moment. Our solar system contains two kinds of planets: the solid rocky ones (Mercury, Venus, Earth, and Mars) and the gas giants (Jupiter, Saturn, Uranus, and Neptune). Rocky planets are made of . . . rock. The gas giants are much bigger than the rocky planets and are composed of gases. The outer two planets, Uranus and Neptune, are in a subgroup of gas giants called ice giants; they

MOONS AND OTHER SMALL BODIES

In our solar system, only Mercury and Venus have no moons. We have just the one; Mars has two small ones (Phobos and Deimos); but Jupiter, the biggest planet, has a whopping 63 moons (though only four are of a comparable size to our moon). Saturn has 61 moons, of which 60 are dwarfed by the huge Titan, which is nearly twice the size of our moon. Uranus has 27 moons and Neptune just 13.

Our solar system also contains quite a lot of extra matter in the form of asteroids or planetoids (which is often used to refer to larger asteroids). These are small lumps of rock that orbit the sun, mostly in a belt between Mars and Jupiter, but they are large enough to do very significant damage to any planet that gets in their way. The rocky Pluto, farther out than Neptune, was once designated a ninth planet but is now considered a large planetoid. Smaller rocks are called meteoroids. Comets are collections of icy and rocky debris, usually with a "tail" of gas and dust streaming out behind them.

---

are less massive than the true gas giants and contain more solid material.

Stars

Stars are made of plasma. Not the same stuff that's in your blood, but a gas that consists partly of charged particles—both positive and negative. In stars, plasma is held together as a big ball by the force of gravity. In the case of our sun and many other stars, the center of the ball is undergoing constant nuclear fusion reactions, whereby hydrogen gas is turned into helium. The energy this generates is radiated out from the star in the form of light and heat.

Every star has a lifespan that begins with its formation from a cloud of mostly hydrogen, and then it passes through the fusion stage. What happens next depends on the size of the star. Medium-sized yellow dwarf stars like our sun will, on finishing up their hydrogen supplies, turn into red giants, their atmospheres expanding massively as their cores collapse under their own gravity. Our sun will do this in about 5 billion years. So if we manage to dodge all the other potential catastrophes out there, we will need to find a new solar system in which to live in at least 3 billion years. The red giant stage proceeds to the white dwarf end stage, when the star sheds its atmosphere and becomes a very dense and much smaller and dimmer body.

DID YOU KNOW?

Scientists believe there are approximately 50 billion galaxies in the universe. They have also discovered at least 20 planets outside our solar system.

If the star is too small for the nuclear fusion process to begin, it's a brown dwarf, or a failed star. Stars intermediate in size between brown dwarfs and yellow dwarfs are red dwarfs, and they live billions of years longer than yellow dwarfs before becoming white dwarfs without passing through a red giant stage.

Very massive stars tend to live fast and die young, in a spectacular gravitational collapse producing a **supernova**—an extremely powerful explosion. A white dwarf star can also go supernova if it draws in enough extra matter. What's left may then become a very hot and very dense but stable neutron star—the biggest become black holes, which are anything but stable. **Black holes** draw in everything in their vicinity, like a whirlpool, and not even light can escape them. There is speculation that there is a supermassive black hole at the center of our galaxy, and indeed at the centers of galaxies generally.

Gravity is what holds together the individual objects in the galaxy and also what keeps moons orbiting planets, planets orbiting stars, and galaxies spinning around their centers. Objects stay in orbit around bigger objects if they are moving fast enough that their forward motion is in balance with the gravitational pull of the bigger object. In the vacuum of space, there's no drag from an atmosphere to slow the object down, so its speed stays constant.

Origins of the Universe

Perhaps the biggest question science has to answer is where the universe came from in the first place. It's easy to see how early civilizations thought it was all about us. Before we knew the nature of the sun and moon and the thousands of stars across the night sky, why would we *not* think our own world was the biggest, most important object in existence and the center of everything? Now we know better, of course, and have come to terms with the fact that we are specks living on a speck orbiting a small star in the corner of a colossal galaxy of billions of such stars, among billions of other galaxies. And in between all that stuff are mind-bending quantities of empty space.

Expanding Universe

Scientists looking at distant stars have noticed something called the **redshift.** The light from distant stars is shifting to the "slower" red end of the visible color spectrum rather than the blue end. This means that the stars are moving away from us and each other. It's the same phenomenon that makes police sirens rise in pitch as they approach us and then lower in pitch when they (hopefully) go past and off into the distance—the **Doppler effect.**

This is the main evidence for the idea that the universe is currently expanding. That doesn't mean that the actual stars and other bits of matter are getting bigger, but that the space between galaxies is increasing. It's also important to understand that the universe isn't expanding into "empty space"; the universe is (mostly) empty space. That this expansion is happening strongly suggests that the universe was once much smaller and that it had a defined beginning.

The Big Bang

Once there was nothing at all; then there was a big explosion and there was everything. Well, not quite. But that's a widespread (mis)understanding of what's meant by the Big Bang. Here's what the current theory says: About 13.7 billion years ago the universe was in an extremely hot and dense state and had a very small volume—the same as an atom—but was very rapidly expanding and cooling. The idea that an atom-size object could hold all the matter that's in the universe today is, obviously, mind-bending.

What happened *before* the Big Bang, we just don't know. But we do know that what we define as the Big Bang wasn't an explosion; rather, it was a massive and extremely rapid expansion. As things cooled down, matter began to coalesce under the influence of the fundamental forces, with free protons and neutrons coming together to form the nuclei of helium atoms, and hydrogen atoms forming from protons and electrons. The coalescence of gas clouds formed star systems, with planets formed from the heavier elements that are blown off from the outer layers of stars as they burn. *Et voilà,* one universe.

What Is Next?

The universe is expanding now, but will it expand forever? Scientists are divided on this. Some say yes; some say no, it will slow down and reach a steady state; some say it will eventually begin to contract instead and all matter and space will return to its original hot, dense, and tiny state in a Big Crunch.

Laws of Physics

This seems like the right time to summarize some—but not all—of the very numerous laws of physics you may have encountered in school.

Laws of Conservation of Energy and Mass

Newton's three laws of motion talk about the relationship between an object and the energy forces it experiences.

- An object remains at rest, or moves in a straight line, until a force acts upon it.
- The acceleration of the object is proportional to the force causing it. It is expressed as

force = mass (of the object) x acceleration

- When object A applies a force to object B, this action produces an equal and opposite reaction in object A.

Laws Governing the Behavior of Gases

Boyle's law says that the pressure of a gas is inversely proportional to its volume, at constant temperatures. So the less space the gas has to move about in, the more pressurized it is.

This is kind of obvious in theory but doesn't actually happen with perfect predictability in reality.

Charles's law turns this around, saying that at a constant pressure, the volume of a gas is directly proportional to its temperature. Heat things up and the gas volume increases. Like Boyle's law, it doesn't work perfectly in real life.

Avogadro's hypothesis says that two different gases of the same volume and at equal pressures and temperatures contain the same number of molecules. It's a hypothesis rather than a law for now, because we don't yet have a good way to count gas molecules.

Laws of Motion

These equations look at the relationship between an object's acceleration, its starting velocity, its final velocity, the time taken from start to finish, and the distance covered in the process. Let's call these *a* (acceleration), *u* (starting velocity), *v* (final velocity), *t* (time), and *s* (distance traveled). Acceleration is measured as the result of final velocity minus the starting velocity, divided by time.

$$a = \frac{v-u}{t}$$

Distance traveled is the time divided by the average speed, which you work out like this:

$$s = \frac{t\,(u+v)}{2}$$

Waves, Radiation, and Space

A wave is a disturbance that carries energy through space and time. Some types of waves need to travel through matter of some sort (whether solid, liquid, or gas), but others can travel through the vacuum of space.

Waves are easy to understand if we start with a kind that we can see. Take a trip to the shore on a reasonably windy day and watch the waves as they roll in. It looks at first as if great volumes of water are moving from way out to sea toward the shore, but that's not the case. Hopefully there's a gull out on the water or some other floating object. If so, watch it and you'll see that it moves up and down as the waves pass through its position, but it isn't carried forward. The water

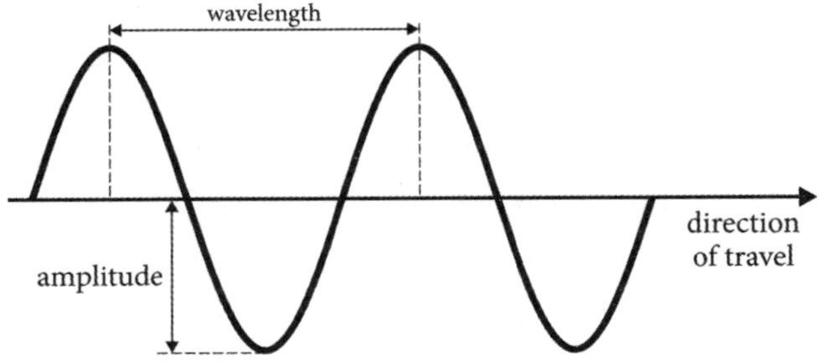

isn't the wave, it's just the material through which the wave passes, and each water molecule moves just a short distance up and down.

So a wave is a pulse or vibration of energy that travels through space and/or through a conducting material of some kind. Waves come in two kinds: **transverse** and **longitudinal.** With transverse waves, the direction of the vibration is at right angles to the direction the wave travels. With longitudinal waves, the vibration and the wave move in the same direction.

To see what a transverse wave looks like, hold a long ribbon or a piece of string by one of its ends and flick it—you'll see the wave travel along to the other end. To see a longitudinal wave, you'll need to dig out your old Slinky toy and lie it flat on the ground. Pull one end out and push it in, and you'll see the wave travel along, the vibration moving at the same angle as the wave itself.

Measuring Waves

There are three measures you can take from a wave—amplitude, wavelength, and frequency. The **amplitude** is how far the wave deviates from its undisturbed path; **wavelength** is the distance from one peak or trough to the next. (The diagram on page 25 shows what this means in a transverse wave); and **frequency** is the number of waves produced per second. Its unit of measurement is the **hertz.** If you know the frequency and wavelength, you can calculate the wave speed using this formula:

wave speed (feet per second) = frequency (hertz) x wavelength (feet)

Sound waves (and also some types of seismic waves, the kind caused by earthquakes) are longitudinal waves. They can't travel through the vacuum of space; they need a conducting material through which to pass, whether it's a solid, liquid, or gas. The amplitude of a sound wave determines how loud the sound is (more amplitude = more volume, hence the term "amplify"). A higher frequency produces a higher-pitched sound. If you keep turning up the frequency, you will end up with sounds too high for human ears to detect, but they will still be audible to some other animals.

Beyond 20,000 hertz (or, as it's more usually written, 20 kilohertz or 20kHz), we get into the range of ultrasound waves. We can't hear these, but their echoes can be used to form images of objects we can't see, like unborn babies.

DID YOU KNOW?

Sound travels four times faster in water than in air.

Light is an example of a transverse wave. Light and other related wavelengths all fall along the electromagnetic spectrum, which is the next topic of discussion.

The Electromagnetic Spectrum

We have already looked at the electromagnetic force. Now it's time to consider the electromagnetic spectrum, which contains all the electromagnetic wavelengths from radio waves at the low-frequency end to gamma radiation at the high-frequency end.

Components of the Spectrum

We give different names to different sections of the spectrum because we have varied uses for them, but it's important to realize that electromagnetic radiation is a continuum—the difference between its components is just a steady increase in frequency, as the simple diagram below illustrates. There's

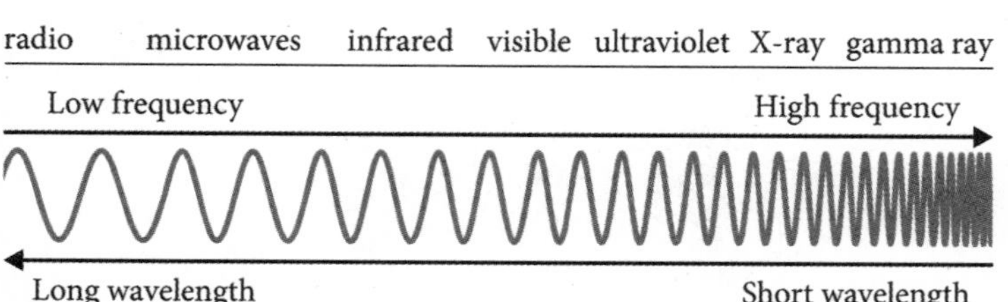

ELECTROMAGNO-WHAT?

The electromagnetic force, as we now know, is the one that's concerned with attraction between charged particles. This force in action also emits—or radiates—waves. The scientist Heinrich Hertz discovered in the late nineteenth century that an electric current generates electromagnetic radiation, and this discovery led to the invention of radio.

In the twentieth century, scientists investigating the nature of electromagnetic waves discovered that they can behave either as waves or as particles called **photons.** This is the founding principle of **quantum mechanics,** which explores the relationship between the nature of an electromagnetic wave and the amount of energy carried by its photons. The simplest equation to express this relationship is:

$$\mathbf{e = h \times f}$$

Here, e = energy, f = frequency, and h is a conversion factor constant (Planck's Constant*).

*The physicist Max Planck discovered that, at a subatomic level, energy can be transferred only in small, discrete units of the same quantity each time. The value of one of these units (or quanta) is 6.626176×10^{-24} joule-seconds. This is called Planck's Constant. Higher-frequency waves carry more energy. Gamma radiation is used to kill cancer cells—that's how fiercely energetic it is.

nothing intrinsically special about the wavelengths of visible light, either; it's just the part we humans happen to be able to see, while other animals can see wavelengths that we can't. Since photons are involved with the entire spectrum, it wouldn't be misleading to describe it all as different wavelengths of light.

Categories of Frequencies

Radio waves occupy the low end of the spectrum, with frequencies up to about 10^4 (i.e., 10 x 10 x 10 x 10) Hz. They are used to carry radio and television signals.

Microwaves come next, their frequencies spanning 10^8–10^{12} Hz. We use them for cooking. They agitate water molecules, and most of the stuff we eat contains a fair bit of water, so confining the food in a sealed chamber and bombarding it with microwaves is an effective way of making it hot. We also use them (at very low power) to carry Bluetooth and Wi-Fi signals.

Infrared waves overlap with microwaves, covering roughly the 10^{11}–10^{14} Hz range. We experience them as heat, and we can use heat-sensitive imaging materials to generate "heat pictures." You might have seen them in action on exciting police chase documentaries, where the infrared cameras can pick out the fleeing criminal as a glowing person-shaped heat source hiding in someone's yard at night.

Visible light comes next, occupying a small band around the 10^{15} mark. "Pure" light combining all visible frequencies is white, but the objects around us absorb some of the wavelengths and reflect others, producing the range of colors we see around us. Shining a beam of white light through a prism onto a sheet of white paper bends the beam and separates out the visible frequencies—we can then see that the colors of light run through red, orange, yellow, green, blue, and violet.

(What happened to indigo? The transitions between colors are arbitrary and subjective, and indigo is often left out.)

Ultraviolet light follows visible light, taking us up to about 10^{17}, though of course ultraviolet light is visible to plenty of animals. We know it as the thing that gives us suntans (and sunburn and potentially skin cancer).

X-rays are around the 10^{18} mark. We know them as the means by which we can make images of stuff (bones) that's underneath other stuff (muscles and skin). X-rays pass through some materials better than others, which is why they are good for looking at bones but not so good at helping us "see" internal organs. X-rays can damage and kill cells, which is why the dentist leaves the room before x-raying your teeth. Occasional exposure to X-rays won't do you any harm, but lots every day will.

Gamma rays have the highest frequency of all at 10^{19} or so, and they are responsible for the radiation we're talking about when we talk about radiation poisoning from spending too much time playing with uranium. Nuclear decay (when the nuclei of atoms break down to make simpler atoms) releases gamma radiation. It's going on in the universe at a tremendous rate, so there is a lot of gamma radiation out there. Luckily our atmosphere absorbs it (though a huge burst of it at close quarters would certainly cause devastation). It can be very harmful to human tissues—although this can sometimes be used to positive effect (to kill cancer cells). Shielding made from very dense materials like lead is the best way to keep out gamma rays.

FICTION

FACT

Radioactive Substances

Naturally occurring radioactive elements are scarce on Earth but can be made in labs. They are elements or forms of elements in which the nuclei have an unstable combination of particles, so they spontaneously break down to form atoms of other simpler and more stable elements—this process is **nuclear decay.** Chemical elements toward the bottom of the periodic table (see page 51) are the most likely to be radioactive.

What Happens in Nuclear Decay?

Here's one of the (many) points where two science disciplines are intimately linked, in this case physics and chemistry. We'll be looking at the structure of atoms in detail in the Chemistry chapter, so here we'll just be touching on those aspects. For now just remember that an atom has a positively charged nucleus surrounded by negatively charged electrons. When a nucleus decays, it releases one or more of three things—an **alpha particle,** a **beta particle,** or a **gamma ray.**

The alpha particle is relatively big and heavy and carries a positive charge. The beta particle is tiny and light and has a negative charge. Both of these particles can damage human tissues, but they are relatively easy to block. The gamma ray has no mass and no charge, but unlike the other two it has a very high penetrative power and gets through most materials. Alpha and gamma radiation are the most dangerous kinds.

Discovery of Radioactivity

In the late 1890s the scientist Marie Curie was investigating the rays emitted by uranium salts when she determined that the radiation was coming directly from the substances' atoms rather than through chemical reactions between atoms. She coined the term radioactivity to describe the effect she discovered and also discovered two new elements, polonium and radium.

Unfortunately, she and her physicist husband, Pierre Curie, both paid a high price for their work, suffering from radiation poisoning because they were unaware at the time how harmful radiation was and had no protection from any

HALF-LIFE

This term describes how long it takes half the atoms of a given quantity of a radioactive substance to undergo nuclear decay. Because we know that the radioactive form of carbon, called carbon-14 (which decays to form nitrogen), has a half-life of 5,700 years, we can calculate the age of a fossil by seeing how much carbon-14 it has compared to its amount of carbon-12—a nonradioactive form of carbon. (Carbon-14 and carbon-12 are examples of isotopes, and they nicely illustrate the point that "isotope" doesn't always mean "radioactive.")

---

of the products of radioactive decay. Marie died in 1934 (her husband had already been killed in an accident) from acute radiation sickness, and even now all of her research notes are too radioactive for safe handling and are kept in lead-lined storage boxes.

Safe Handling of Radioactive Materials

Now that we know how dangerous radiation is, we won't be in any hurry to start juggling chunks of uranium without some sort of special protection. Lead is dense enough to significantly cut down gamma radiation, so it's often used as shielding in labs where radioactive materials are used. Fallout shelters, designed to protect people from radiation after a nuclear bomb attack, are shielded with various materials. If they are underground, a yard of packed earth on the roof is necessary to block the gamma rays.

Safely disposing of nuclear waste from nuclear power stations is a real problem because the spent nuclear fuel is still somewhat radioactive. It needs to be treated to make it as inert as possible and is sometimes stored deep underground, away from where people are living.

Energy and Electricity

Electricity has to be everyone's favorite energy source. It's easy to work with, and we can use it to power all kinds of machines and processes. Naturally occurring electricity is rare and unpredictable—harnessing lightning strikes isn't really a practical option, for example. So we generate our electricity by converting other, more easily obtainable energy sources.

Natural Energy Sources

Our world is full of many different potential sources of energy that we can convert into electricity. Some of them are sustainable—we can keep on using them indefinitely; others aren't—we'll use them up eventually and they can't be replaced.

Fossil fuels are coal, natural gas, and oil. They are called fossil fuels because they are the compressed remains of ancient forests (coal) or tiny sea animals (gas and oil) squashed up and broken down into carbon-based compounds that burn readily, releasing heat. They are not sustainable, because the conditions that produced them no longer exist. Also, burning them releases extra carbon dioxide and other unwelcome gases into Earth's atmosphere, contributing to global warming.

Nuclear power is generated by breaking up atoms of plutonium and uranium. When nuclei of atoms from these two elements are battered with particles called neutrons, they release heat energy. Inside the reactor of a nuclear power station, this process (nuclear fission) is initiated and controlled and the energy harnessed. The process doesn't release toxic gases, but the waste it generates is dangerous. Nuclear fuel is unsustainable, and an accident at a nuclear power station could have very far-reaching and disastrous consequences.

Wind power is a sustainable source of energy. Large wind turbines, like giant futuristic windmills, are sited in exposed places with strong prevailing winds. The wind turns the turbine's blades, and this kinetic (moving) energy is captured. No pollutants are produced, but wind farms can be unsightly, noisy, and dangerous to birds.

Water power is another sustainable energy source. Hydroelectric power stations use dams built across rivers and capture the water's energy as it flows down pipes built into the dam. It doesn't produce pollution, but dams do damage wildlife habitats.

Geothermal power stations capture the natural heat energy from volcanic areas. It is sustainable and does not produce pollutants, but its availability is limited.

Other energy sources include solar power, wood, biofuels (oils derived from crops that can be burned like fossil fuels), and hydrogen fuel cells.

$E = mc^2$

This most famous formula is from Einstein's Theory of Special Relativity and was first proposed in 1905. It just means that matter (any kind of physical substance—like rock, a star, a bar of chocolate, your cat) and energy are different versions of the same thing. This comes into play in nuclear fission: The products of the fission reaction together weigh slightly less than the original fission material because the "missing mass" has been converted into energy.

$$E = mc^2$$

E = energy; m = mass (measured in grams and kilograms); c = the speed of light (which is a constant; i.e., it doesn't change, no matter what).

The formula shows us how much energy is stored in a quantity of matter. The main points to understand from this are that a) matter is a kind of energy, and b) energy doesn't go anywhere. You can't add new energy to the universe, and you can't take any away, but you can change it from one form to another. What this has to do with generating electricity is that we can convert any form of energy into any other form of energy without "losing" any of it. Some of it might escape from our control, but if our energy-converting machinery is well designed, we can keep this to a minimum.

DID YOU KNOW?

In September 2011, scientists at the CERN Laboratory in Switzerland sent some neutrinos through their particle accelerator and discovered that their neutrinos had reached their destination at a speed faster than the speed of light. Previously it had been thought that nothing could go faster than the speed of light. Scientists the world over are studying this experiment and trying to analyze its implications.

Kinetic Energy

Whatever the source of energy, the process of converting it to electricity requires that the energy first be made into kinetic energy (the energy of motion), and then this kinetic energy is used to drive a generator to produce the electricity. In the case of wind and water power, the harnessed energy is kinetic already. With the others, heat energy must be converted to kinetic energy—usually by using the heat to turn water into steam, which then powers a turbine of some kind. The generator has an electric charge in its workings already, but it's the input of kinetic energy that causes the electric charge to flow.

Heat Transfer and Efficiency

Moving heat energy around in the most efficient way possible is a major human preoccupation; if we'd never learned how to do it, we wouldn't have been able to colonize such a large proportion of the available land surface on Earth. Heat has three ways of getting around: conduction, convection, and radiation.

Conduction. Heat can travel through matter in any form: solid, liquid, or gas. If you hold a spoon over a candle flame, pretty soon that spoon will get too hot to hold. Some materials—like metals—conduct heat better than others.

Convection. Heat moves through liquids and gases by convection. When you add extra hot water to your bath, the heated water expands (or, to be precise, the spaces between the water molecules get bigger as the molecules start moving around more), becoming less dense and thus taking up more space. It rises above the more dense cooler water, which sinks. The rising of hotter, less dense liquids or gases and falling of cooler and more dense liquids or gases produces convection currents. You might see birds of prey taking advantage of the rising hot-air currents—thermals—that form by hillsides to gain height effortlessly. So convection describes the tendency of heat to go up and is the reason we don't put our room heaters on the ceiling.

Radiation. Unlike conduction and convection, which describe the transfer of heat through the particles that make up matter, radiation comes in waves. It's a part of the electromagnetic spectrum that describes all kinds of waves (see page 28). Infrared radiation = heat. Every kind of matter absorbs and gives off (reflects) infrared radiation. Things with more surface area for their volume (long and narrow) absorb and reflect more than things with less (short and wide). Objects that are white and/or shiny reflect more than those that are dark and/or matte, which absorb more.

All of these three modes of transport can be used to heat up the rooms in our homes but need to be blocked as much as possible to keep the heat in where we want it, rather than letting it escape into the wider world. Those who live in the tropics have the same problem in reverse—outside heat needs to be kept out if it's too excessive. For this reason we

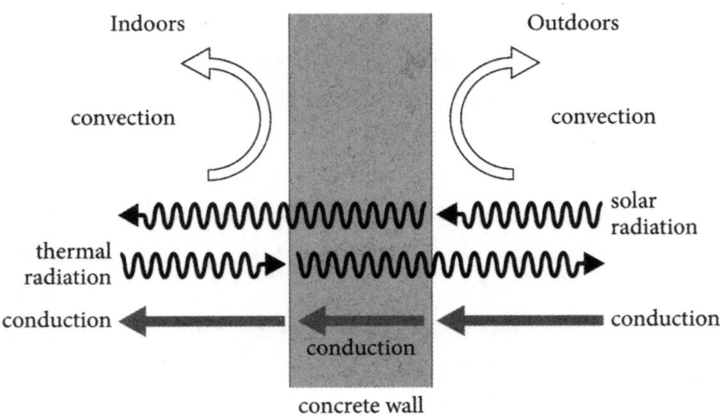

need to use **insulation**—materials that discourage conduction, convection, and radiation through our walls, windows, and roofs.

Because of convection, buildings lose the most heat through the roof. All the materials that the building's outer shell is made from can conduct heat out of (or into) the building. Therefore, we need to use materials that are good

THERMODYNAMICS

Time to lay down the law. Or laws, in this case. The movement of heat is governed by four physical laws, which are as follows (converted into sensible language).

The Zeroth Law: If two heat systems are both in equilibrium with a third, they are also in equilibrium with each other. (It's numbered "zeroth" because it was added as an afterthought, being almost too obvious to be worth writing down.)

The First Law: In a closed system, the amount of heat remains constant.

The Second Law: In a closed system, heat will always spread out as evenly as it can. (Another way to put this is that its entropy, or state of disorderedness, will always increase.)

The Third Law: At the temperature absolute zero, the entropy of a system is zero.

The Second Law is the one of interest here, because it's what we are working against when we try to keep our homes warmer (or cooler) than the world around us.

insulators (bad conductors, in other words) when we make our buildings. Fluffy fiberglass wall insulation is a good example—it is a poor conductor, and it also traps pockets of air. Still air is a poor conductor, too, and by preventing air circulation, you also reduce heat loss by convection. Double glazing windows is another way of trapping air, and blocking drafts has a similar effect.

Using Electricity

Electricity is immensely useful but also potentially dangerous. To be used safely in our homes, it needs to be at an appropriate **voltage.** However, large-scale transportation of electricity requires a much higher voltage, which is itself lower than that of electricity fresh from the power station generator.

Measurements of Electricity

Electric current is measured in **amperes** (amps for short)—an expression of the amount of current flowing through a conductor over a specified unit of time. A **volt** is a measurement of potential difference—the difference in electric charge between two points. The negatively charged particles—**electrons**—that make up an electric current flow from areas of high negative charge to areas of low negative charge (in other words, areas of high positive charge). Voltage can be calculated by dividing power (in watts) by amps. The formula looks like this:

volts = watts/amps

or this:

watts = amps x volts

or this:

amps = watts/volts

To help get your head around this, imagine water flowing through a hose attached to a tap. The rate of flow (amps) can be made more powerful (watts) by turning on the tap more fully to put more pressure on the water (volts).

Changing Voltages

To move the electricity from power stations to homes, we need to give it a low current to minimize energy loss—high currents running through wires lose energy in the form of heat. A low current means a high voltage. We use transformers to produce the high voltages necessary. Because such voltages are very dangerous, the design and siting of pylons and power lines needs to be planned very carefully.

When the electricity goes into our homes, the voltage has to be dropped to a safer level, so transformers come into play again. These are called step-down transformers—the ones that increase voltage are step-up transformers. Both kinds use magnetic fields to alter voltage.

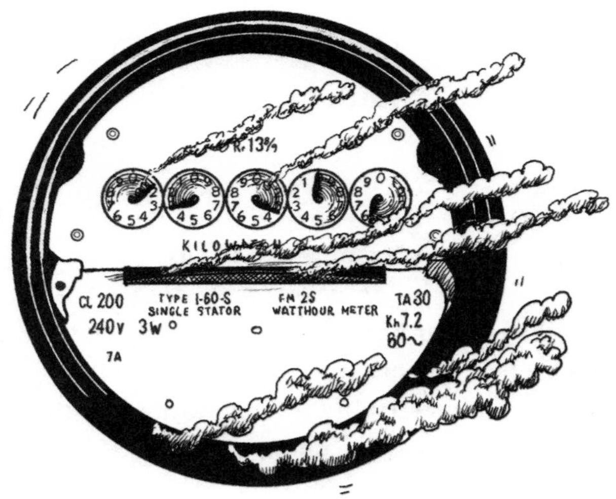

Resistance

Resistance isn't futile when you're talking about electricity—far from it. Lots of things can impede the progress of an electric current, and sometimes we purposely use resistors because they change voltages. The equation looks like this:

voltage = current x resistance (measured in ohms)

In other words, when a current meets resistance, it needs more voltage to overcome that resistance.

Using Electricity in Your Home

We know that different appliances use different amounts of power and affect the size of our bills. The energy an appliance uses is measured in units called **kilowatt hours (kWh).** It's calculated by this formula:

energy used (kWh) = power (kilowatts) x time (hours)

If you also know the cost of your electricity per unit, you can work out how much it costs to run each appliance for an hour by multiplying the number of kilowatt hours by the cost per unit. Of course, appliances use different amounts of energy in different modes—a laptop uses more power when in use than when on standby, for example, so for a really accurate calculation you'd need to factor that in, too.

2

Chemistry

The Periodic Table

The periodic table lists all known **chemical elements** in order of their **atomic number** (the number of protons in their atoms). The table is displayed in such a way that elements with similar and recurring (or periodic) properties form columns. This is convenient for understanding periodic properties but does produce a shape with large gaps between the left and right sides of the table in the first three rows.

The 10 columns that unite the two sides below the first three rows contain the transition elements, which bridge the gap between metals and nonmetals. Two rows from the lower section (the lanthanides and the actinides) are usually extracted and shown separately at the bottom because they have properties in common as you read across rather than down.

What Is a Chemical Element?

A chemical element is a substance that is pure; it can't be broken down by any chemical process into another substance, and all of its atoms have the same structure. Most elements are naturally unstable in the prevailing conditions here on planet Earth, and they are inclined to react with one another to form stable compounds that are nothing like their elemental

THE PERIODIC TABLE OF THE ELEMENTS

1 H Hydrogen 1.00794																	2 He Helium 4.003
3 Li Lithium 6.941	4 Be Beryllium 9.012182											5 B Boron 10.811	6 C Carbon 12.0107	7 N Nitrogen 14.00674	8 O Oxygen 15.9994	9 F Fluorine 18.9984032	10 Ne Neon 20.1797
11 Na Sodium 23	12 Mg Magnesium 24											13 Al Aluminium 26.981538	14 Si Silicon 28.0855	15 P Phosphorus 30.973761	16 S Sulphur 32.066	17 Cl Chlorine 35.4527	18 Ar Argon 39.948
19 K Potassium 39.0983	20 Ca Calcium 40.078	21 Sc Scandium 44.955910	22 Ti Titanium 47.867	23 V Vanadium 50.9415	24 Cr Chromium 51.9961	25 Mn Manganese 54.938049	26 Fe Iron 55.845	27 Co Cobalt 58.933200	28 Ni Nickel 58.6934	29 Cu Copper 63.546	30 Zn Zinc 65.39	31 Ga Gallium 69.723	32 Ge Germanium 72.61	33 As Arsenic 74.92160	34 Se Selenium 78.96	35 Br Bromine 79.904	36 Kr Krypton 83.80
37 Rb Rubidium 85.4678	38 Sr Strontium 87.62	39 Y Yttrium 88.90585	40 Zr Zirconium 91.224	41 Nb Niobium 92.90638	42 Mo Molybdenum 95.94	43 Tc Technetium (98)	44 Ru Ruthenium 101.07	45 Rh Rhodium 102.90550	46 Pd Palladium 106.42	47 Ag Silver 107.8682	48 Cd Cadmium 112.411	49 In Indium 114.818	50 Sn Tin 118.710	51 Sb Antimony 121.760	52 Te Tellurium 127.60	53 I Iodine 126.90447	54 Xe Xenon 131.29
55 Cs Cesium 132.90545	56 Ba Barium 137.327	57 La Lanthanum 138.9055	72 Hf Hafnium 178.49	73 Ta Tantalum 180.9479	74 W Tungsten 183.84	75 Re Rhenium 186.207	76 Os Osmium 190.23	77 Ir Iridium 192.217	78 Pt Platinum 195.078	79 Au Gold 196.96655	80 Hg Mercury 200.59	81 Tl Thallium 204.3833	82 Pb Lead 207.2	83 Bi Bismuth 208.98038	84 Po Polonium (209)	85 At Astatine (210)	86 Rn Radon (222)
87 Fr Francium (223)	88 Ra Radium (226)	89 Ac Actinium (227)	104 Rf Rutherfordium (267)	105 Db Dubnium (268)	106 Sg Seaborgium (271)	107 Bh Bohrium (272)	108 Hs Hassium (270)	109 Mt Meitnerium (276)	110 Ds Darmstadtium (281)	111 Rg Roentgenium (280)	112 Uub Ununbium (285)	113 Uut Ununtrium (284)	114 Uuq Ununquadium (289)	115 Uup Ununpentium (288)	116 Uuh Ununhexium (293)	117 Uus Ununseptium (294)	118 Uuo Ununoctium (294)

58 Ce Cerium 140.116	59 Pr Praseodymium 140.90765	60 Nd Neodymium 144.24	61 Pm Promethium (145)	62 Sm Samarium 150.36	63 Eu Europium 151.964	64 Gd Gadolinium 157.25	65 Tb Terbium 158.92534	66 Dy Dysprosium 162.50	67 Ho Holmium 164.93032	68 Er Erbium 167.26	69 Tm Thulium 168.93421	70 Yb Ytterbium 173.04	71 Lu Lutetium 174.967
90 Th Thorium 232.0381	91 Pa Protactinium 231.03588	92 U Uranium 238.0289	93 Np Neptunium (237)	94 Pu Plutonium (244)	95 Am Americium (243)	96 Cm Curium (247)	97 Bk Berkelium (247)	98 Cf Californium (251)	99 Es Einsteinium (252)	100 Fm Fermium (257)	101 Md Mendelevium (258)	102 No Nobelium (259)	103 Lr Lawrencium (262)

CHEMICAL NICKNAMES

Each element has a shortened name, which is often used in displays of the periodic table and always in chemistry equations. Most of them are fairly obvious shortenings of the full name, like H for hydrogen, He for helium, Al for aluminum, and so on. Where this is not the case, it is generally because there is another language involved. For example, Au comes from *aurum,* the Latin word for gold. Na for sodium also comes from its Latin name (*natrium*). All the nicknames are one or two letters long, apart from those for ununbium (Uub), ununhexium (Uuh), ununoctium (Uuo), ununpentium (Uup), ununquadium (Uuq), ununseptium (Uus), and ununtrium (Uut)—and, frankly, the less said about them the better.

---

components. For example, you probably won't encounter chlorine, a smelly and poisonous green gas, or sodium, a squishy metal that fizzes furiously in contact with water, in their natural state anywhere outside a chemistry laboratory. However, you will find the stable compound formed between them—sodium chloride, aka "table salt"—in your kitchen cupboard and many other places besides. (Lots of things are "salts" in chemistry, but not all of them are delicious on chips.)

When the Russian scientist Dimitri Mendeleyev was first putting together the periodic table, there were many gaps, but the predictable repeated patterns of element properties meant that he could spot where an element was missing. This also meant that it was possible to make a very good guess as to what the missing element would be like and how to find it—and that process continues today. No element above the atomic number 92 exists in nature (on this planet, anyway), but elements have been created in particle accelerators.

Some Element Groups

Columns of elements in the periodic table that show periodic properties are called groups. Let's take a look at some of the most distinctive groups.

Alkali Metals. The first group are soft and very reactive metals that oxidize (react with oxygen) quickly. In the lab they are normally kept in oil until it's time to cut a piece off and throw it into a dish of water to demonstrate how reactive it is. The reaction becomes increasingly violent as you move down the group, with lithium at the top fizzing around gently, and cesium near the bottom producing a violent explosion.

Alkali Earth Metals. Soft and reactive, though less so than the alkali metals.

Halogens. These start out at the top of the column as gases (fluorine, chlorine), then liquid (bromine), and finally solids (iodine, and astatine—the rarest naturally occurring element, of which there are about 30 grams in the entire crust of the Earth). Halogens become less reactive as you move down the group, in contrast to alkali metals, but all are dangerous in their pure form. Fluorine is especially nasty (though good for the teeth). Most nonmetals burst into flame in its presence, and it reacts even with something as innocuous as glass.

Noble Gases. The very epitome of inertness, noble gases react with almost nothing. This is because each electron shell around their atoms has exactly the right number of electrons (see Chemical bonds on page 59 for more about electron shells). They include helium, neon, and argon, and while all are gases, they get progressively heavier as you move down the group. For example, a balloon filled with radon will plummet like a stone.

Atomic Structure

A single atom is the smallest amount of an element that you can get. Each atom is made up of equal numbers of positively charged **protons** and negatively charged **electrons.** Therefore, the atom as a whole has no electric charge. Two or more atoms (whether of the same element or two or more different ones) bonded together become a **molecule.**

The protons form a central **nucleus,** while the much smaller electrons orbit this nucleus in shells or layers, progressively farther out from the nucleus, with each shell having an increased maximum capacity of electrons. The nucleus contains variable numbers of a third particle type, the **neutron,** which has no electric charge. There are usually as

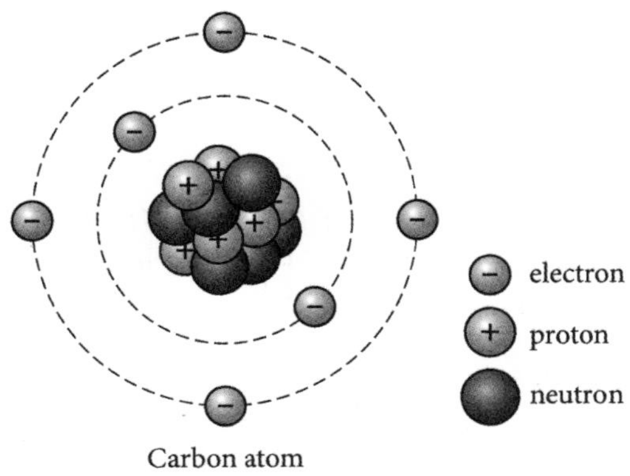

Carbon atom

many neutrons as there are protons, but not always—and the smallest atom of all, hydrogen, usually has no neutrons. Look at the diagram of a carbon atom, showing how the electrons are arranged in their shells around the nucleus.

An element's atomic number (the number that appears above its abbreviated name in most periodic table layouts) tells you how many protons it has in each atom (and therefore, by extension, how many electrons—under normal circumstances, at least).

Atomic Number and Atomic Mass

In some periodic tables, you'll see two numbers—the atomic number, which progresses in single whole numbers from 1 for hydrogen, 2 for helium, 3 for lithium and so on, and the atomic mass, which is usually about double that of the atomic number at the top of the table but is getting to three times the atomic number down at the bottom. The atomic mass isn't always shown as a whole number. For example, in the table on page 51, fluorine's atomic mass is 18.9884032.

We already know that the atomic number of an element denotes the number of protons in one atom. Atomic mass can be considered as the number of protons plus neutrons. (Actually, the more accurate definition is more complex than that, but for our purposes this definition works quite well.)

Obviously, for an individual atom, the atomic mass will always be a whole number. A carbon atom contains six protons and six neutrons, so its atomic mass is 12. But wait, not every carbon atom has six neutrons; most do—in fact, 98.89 percent do in natural conditions. But the others have either one or

ANIONS AND CATIONS

If an atom has more or fewer electrons than it has protons, then it is an ion, with an electric charge. More electrons than protons = negative charge = anion. Fewer electrons than protons = positive charge = cation. Molecules with missing or extra electrons are ions, too.

So how might an atom lose or gain an electron? Usually this happens during chemical reactions (of which more later). Here on Earth most ions don't hang around long in normal outdoor environments, because they are inherently unstable. But we can make them artificially by applying a lot of the right kind of radiation in controlled conditions. We also make them without thinking about it inside our own bodies, where they do vital and useful things like allowing our nerve cells to send electrical impulses. Ions are important in many natural and industrial processes, and stars are made of ionized gas (plasma).

All of which gives rise to the best (if not the only) chemistry joke of all time. Two hydrogen atoms are walking down the road, when one stops dead with a look of great anxiety. The other says, "Hey, what's wrong?" The first says, "I lost my electron!" The second says, "Oh, no! Are you sure?" The first replies, "Yes, I'm positive."

---

two extra neutrons. The three varieties are called **isotopes.** They are specified by their atomic mass—the normal carbon isotope with an atomic mass of 12 is called carbon-12, the isotope with one extra neutron is carbon-13, and the one with two extras is carbon-14. The first two are stable, but carbon-14 is not—it's formed in the atmosphere by cosmic radiation but is radioactive and turns quickly into nitrogen.

The reason, therefore, that these values in periodic tables are not whole numbers is that they proportionately express the average atomic weight of all the known isotopes—this average is called the atomic weight, or relative atomic mass. In carbon's case, that's 12.0107.

Even if you have limited knowledge about nuclear power, you'll probably put "isotopes," "radioactivity," and "nuclear power" in the same part of your memory. Radioactivity and nuclear reactions happen because of the behavior of the neutrons within atoms of unstable isotopes.

Chemical Bonds

Remember the carbon atom on page 55? It has six of everything—six neutrons and six protons in the nucleus, and six electrons orbiting that nucleus. The electrons are arranged in an inner shell of two and an outer shell of four.

So far, so good. But every electron shell has a maximum number of electrons that it can accommodate, and the second shell can hold up to eight of them. In the case of a carbon atom, there are "vacancies" in that outer shell for four more electrons. Therefore, each carbon atom is unstable on its own and "wants" to combine with other atoms—whether they are of carbon or a different element—to form a molecule whose constituent atoms each have a "full" outer shell containing eight electrons, some of which are shared between the atoms. Such combining of atoms is a **chemical reaction.**

Hydrogen and Helium

The simplest example of how this works comes from the simplest element—hydrogen. It has only one electron, but it "wants" to have two, because the innermost electron shell is only complete with two. Therefore, hydrogen in its natural state comes in molecules of two atoms (H_2), which share their two electrons, so each has a full shell. They can be represented like this:

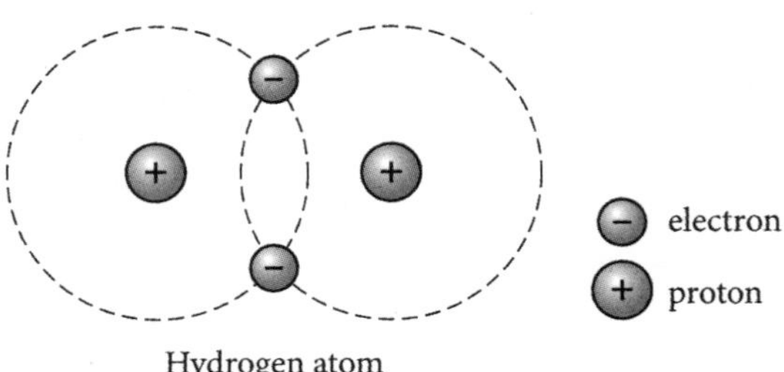

Hydrogen atom

Helium has two electrons per atom. That completes the electron shell without the need for any combining. Therefore, helium naturally exists as individual atoms and is disinclined to react with other substances. All of the other noble gases have complete electron shells, too, so are similarly unreactive. But most other elements have incomplete electron shells, so they don't occur in their natural state as individual atoms but as multi-atom molecules.

Here's another example: the carbon dioxide molecule, formed from one carbon and two oxygen atoms. Each oxygen atom shares two of its outer-shell electrons with the carbon atom, and the carbon atom shares all four of its outer-shell electrons, so all three atoms have a full outer shell of eight electrons.

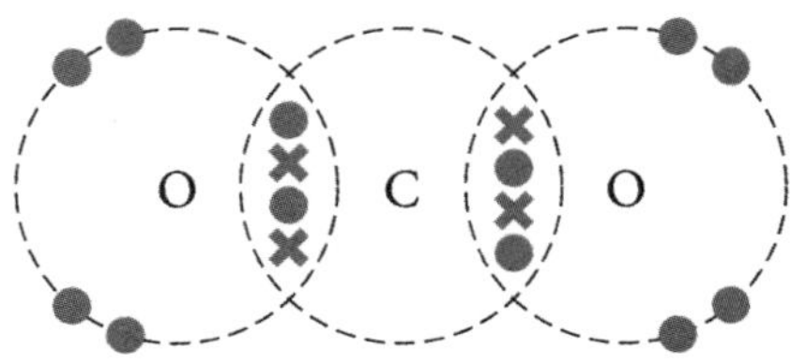

Types of Bond

When atoms share electrons like this, the bond formed is called a **covalent bond.** It is generally a very strong chemical bond, especially if two or more electrons are shared. Covalent bonds tend to be weaker when they involve electron shells that are farther out from the nucleus. The **valency** of an element (the number of chemical bonds formed by its atoms) tells you how many electrons it "needs" to make a compound. Hydrogen has a valency of 1, oxygen's is 2; therefore, their compound—water—is H_2O.

Ionic bonds are very strong bonds formed between metals and nonmetals, when each has either lost or gained an electron to become an ion (see Anions and Cations, page 57). While covalent bonds involve sharing electrons, ionic bonds involve exchanging them. Sodium and chlorine form an ionic bond when sodium loses its single outer shell atom to become a positively charged ion with two full electron shells, and chlorine takes up that electron to become a negatively charged ion with three full electron shells. In case you tuned out during that last sentence, hopefully the illustration on the following page will make things clearer.

There are several kinds of bonds between molecules. They are what keep a solid substance like ice, carbon, or titanium held together, often in a regular, crystalline arrangement, but in general the bonds are much weaker than the ones that form between atoms. One of the most common is the **hydrogen bond,** which is formed by the attraction between the positively charged nucleus of a hydrogen atom within one molecule, and a different kind of atom in another molecule. Hydrogen

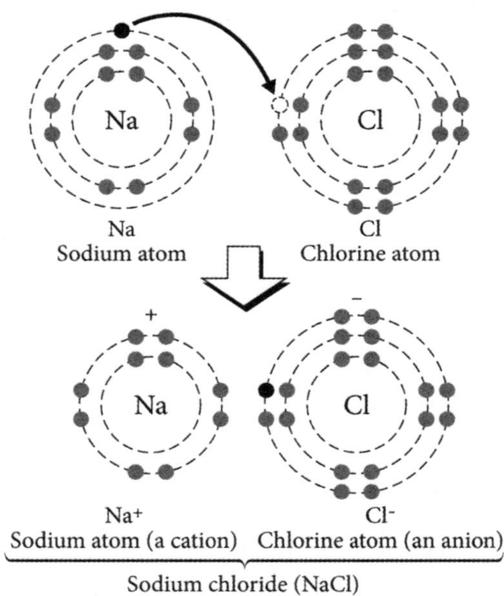

bonds are what hold water molecules firmly together in ice and loosely together in liquid water.

The **metallic bond** is what holds the atoms of a metal together. It's like a covalent bond in that electrons are shared, but they are shared between all atoms at once, producing what's sometimes called a "sea of electrons" shared freely among the nuclei of the metal atoms—the structure is sometimes called a metallic matrix. Therefore, it makes more sense to refer to "metallic bonding" because there is no single individual bond.

ELECTRONEGATIVITY

Sharing electrons sounds very fair-minded, but some kinds of atoms are less willing to share than others. Let's consider that most volatile of elements: fluorine. Fluorine atoms pull electrons toward themselves more strongly than any other kind; they have the highest electronegativity. So in a molecule of hydrogen fluoride, the two electrons that form the bond between them are always closer to the fluoride end, giving that end of the molecule a negative charge and the hydrogen end a positive charge. Molecules like this are called **polar molecules.**

This type of bonding is what makes metals ductile (stretchy) and/or malleable (squashy rather than brittle).

Solids, Liquids, and Gases

Bonds, whatever their type, are what determine the **state,** or phase, of a substance—whether it is solid, liquid, or gas. Let's look at a nice ordinary substance, say, water. Fresh from the freezer it is a solid. Its atoms connect up into regular crystals, which are joined to each other in an equally regular pattern. In ice the crystals are like hexagonal wheels. In other solids they may be different in shape, but all true solids have a regular crystalline structure of some kind or another. Those that don't, like glass, are really just very slow-flowing liquids.

After a while at room temperature (specifically, above 32°F/0°C), the ice crystals will melt. That energy is enough to start breaking down the hydrogen bonds that hold the

structure together, so a phase change occurs from solid to liquid. Put your water into a pan on the stove and turn it on, and it will boil—the hydrogen bonds all completely break down and the water becomes a free gas (or steam).

Chemical Reactions

We've already seen why elements react together to form compounds—it has to do with going from an unstable state with incomplete electron shells to a stable one with complete shells. Compounds may also react together for the same reasons, because some compounds have a more stable structure than others. The chemical bonds that hold the atoms of a compound together are generally strong and under normal circumstances aren't readily broken. For that reason, we don't observe many rapid, dramatic chemical reactions going on in the world outside, although there are any number of them going on all the time inside us.

Rust to Rust

Any reaction between a metal and a gas is called **oxidation,** though the term is most commonly used (for obvious reasons) when the gas in question is oxygen.

They say a watched pot never boils. It's not true; it's just very boring waiting for it to happen. Even more boring is waiting for the pot to become rusty. We tend to think of chemical reactions as rapid and dramatic, but the reaction between iron and oxygen is slow and gradual. The resulting crumbly orange-red stuff is an iron oxide, a compound of

THE MATHEMATICS OF CHEMISTRY

Chemical reactions are usually shown as equations, with the reactants (things that are reacting) on the left and the products (what they turn into) on the right, joined by an arrow. Here's an example:

$$CH_4 + 2\ O_2 \longrightarrow CO_2 + 2\ H_2O$$

This equation is what happens when you burn methane gas. Every molecule of methane (CH_4) reacts with two molecules of oxygen (O_2) to produce one molecule of carbon dioxide (CO_2) and two molecules of water (H_2O).

Equations involving charged atoms (ions) may show the charges on the atoms involved. For example, this is the reaction between silver (Ag) and chloride (Cl) ions:

$$Ag^+ + Cl^- = AgCl$$

---

iron and oxygen with ionic bonds. Water is necessary to make rusting occur because it acts as an **electrolyte**—helping the free electrons released from the iron atoms to unite with the oxygen atoms.

The process is indeed slow by our standards but has been going on across planet Earth for a long time, which is why there's very little pure iron to be found naturally anywhere. Therefore, it's only man-made iron things that we observe becoming rusty. So where did we get the iron from to make the pot, the bicycle, and all that other rusty stuff? How do we turn iron oxide back into iron?

First, you need some iron oxide. This is found naturally in the form of iron ore, rocks rich in iron oxides. They tend to have a giveaway reddish color. Heating up iron ore in the presence of carbon eventually causes the oxygen to separate from the iron and instead bond with the carbon, which it does readily because carbon-oxygen bonds are stronger than iron-oxygen ones. The result is the production of carbon dioxide and pure molten iron.

Energy

You often have to put in lots of energy to bring about a chemical reaction. When you take ice out of the freezer, the energy from its new warmer surroundings starts to break the hydrogen bonds that hold the molecules together, freeing the molecules to move around more so it becomes liquid. More heat and more bonds are broken, until you end up with a gas in which the molecules aren't bonded together at all. It takes a whole lot more energy to crack the bonds between the atoms themselves.

Set fire to a piece of wood, and the heat starts to break down the various hydrocarbon compounds that it's made of, releasing hydrogen and carbon, which reacts with oxygen in the air to produce smoke. By the time you've finished, what's left is a black lump of almost pure carbon.

So it takes energy to break up compounds, but when compounds are formed, energy is released. Think back to that lump of sodium in the dish of water. As the sodium combines with the hydrogen and oxygen in the water to produce sodium hydroxide, lots of energy is generated, making the sodium lump fizz and bounce around on the water's surface. If your sodium lump is too big, you'll get an explosion.

ACID TEST

An **acid** is a compound that when dissolved in water produces hydrogen ions. An alkali (or, to give it its more correct name, a **base**) is something that, when dissolved, accepts hydrogen ions. The **pH** (potential of hydrogen) scale expresses how acidic something is—pure water has a pH of 7, which is the midpoint; acids come in at under 7; and bases over 7. When you put an acid and a base together, you get a reaction that produces a salt of some kind, with water as a by-product. One way of making everyone's favorite salt, sodium chloride, is to combine dissolved sodium hydroxide with hydrochloric acid. Then you just get it warm enough for the water to evaporate and you'll be left with solid sodium chloride crystals.

Collision Theory and Rates of Reaction

So we know that some chemicals react with others under the right conditions. We can also observe that the reaction between the same two chemicals may be faster or slower, depending on a variety of external factors. To explain exactly how reactions occur and why their speed may vary, chemists Max Trautz and William Lewis came up with the **collision theory.**

The key premise to collision theory is simple: To react, two molecules or atoms must literally collide with sufficient speed and at the right angle so that existing chemical bonds are broken. Suitable collisions will happen more often when

the chemicals are at a higher concentration, and also under higher temperatures, when the atoms/molecules are moving about more rapidly.

Angles of Interest

If the molecules don't collide at the right angle, there won't be a reaction. Bonds need to be broken for a reaction to occur, and bonds (being formed by shared electrons) have a negative charge, so it takes a positive charge to break them. In the last section, we looked at the molecule hydrogen fluoride, which is negatively charged at its fluoride end and positively charged at the hydrogen end. To break a bond in another molecule and cause a reaction, the hydrogen end of the molecule has to hit the bond. If the fluoride end hits, it's negative-meets-negative and the two repel each other.

Reaction Rates

A gentle collision doesn't break anything. A certain amount of energy is needed before anything interesting will happen—what is known as the **activation energy.** How much this is varies from reaction to reaction, but once it's been achieved, a reaction will start to occur. How quickly the reaction happens

CONCENTRATION SPAN

When you throw a handful of table salt into some water, the salt dissolves and you get salty water—a solution of sodium chloride. The sodium chloride doesn't actually react with the water to form a different compound, but its crystals completely break down into individual sodium and chloride ions, forming a homogeneous (uniform) mixture. In this case the salt is the **solute** (the thing that dissolves) and the water is the **solvent** (the thing into which it dissolves).

The concentration of a solution is defined as the number of moles of solute per liter of the solution. A **mole** is a unit of quantity in chemistry. One mole of anything is the amount that contains the same number of particles as there are in 12 grams of carbon of the carbon-12 isotope. Why this figure, and why the name "moles," is anyone's guess. We think of solutions as being solids dissolved in liquids, but you can also have solutions of gases dissolved in other gases (the air we breathe is an example—various gases dissolved in nitrogen), solids dissolved in other solids (as with stainless steel, in which carbon atoms are mixed in with the crystals of iron), or other combinations. The relative amounts of solute and solvent determine the concentration of the solution. The spread of molecules of solute through the solvent is called **diffusion.** The tendency for solute molecules to move from areas of high concentration to areas of low concentration is called **osmosis.**

now is determined by the concentration of the two things reacting together, as well as various other factors; it can be predicted by horribly complex-looking equations.

Catalysts

Catalyst is a chemistry term that has found its way into general parlance without really changing its meaning. In chemistry a catalyst is something that causes a reaction rate to increase—it "helps things along." Catalysts are much used in industry to speed up chemical reactions. Usually what happens is that the catalyst first reacts readily with one of the reactants to form a compound that then combines with the other reactant. The unstable result then quickly turns into the expected product plus the catalyst reconstituted. We can show this as a sequence. Let's say A is the first reactant, B the second reactant, P the product of a reaction between them, and C a catalyst.

{without catalyst}	**A + B —> P**
{with catalyst}	**A + C —> AC**
	AC + B —>ABC
	ABC —> CP
	CP —> C + P

The route with the catalyst looks more convoluted, but because all the stages happen much more quickly than A + B —> P, the overall reaction rate is increased. A huge variety of substances can work as catalysts under a number of different conditions.

Air, Fuels, and Pollution

Here on planet Earth, the air we breathe is a mixture of various gases:

Nitrogen. By volume air is a little over 78 percent nitrogen. Nitrogen forms molecules of two atoms each, joined by a triple covalent bond (formed of six electrons, three from each atom). This bond is extremely strong, making N_2 very nonreactive. That's good news for us, because it means we can breathe it in and out without it reacting with anything inside our bodies, and it doesn't readily react with other things that we might accidentally or purposely let loose into Earth's atmosphere.

Oxygen. The next most abundant gas in our air is oxygen, making up nearly 21 percent. Oxygen also forms two-atom molecules, but with a single rather than a double bond, making oxygen more reactive. Animals consume oxygen, but luckily plants produce it.

The Final 1 Percent. The remaining 1 percent of air is composed mostly of argon, one of the noble or inert gases. Other noble gases put in an appearance, too, with helium, neon, krypton, and xenon contributing trace amounts. Carbon dioxide makes up about 0.03 percent, and there are tiny amounts of hydrogen, methane, carbon monoxide, and assorted others.

Water Vapor. Water vapor is usually excluded from breakdowns of air composition, because its presence is extremely variable, given its tendency to turn into rain or snow and fall down, only to evaporate under the sun and rise up again. It can make up as much as 4 percent of the atmosphere (air that is 4 percent water vapor is at 100 percent humidity).

Air Pollution

So much for what *should* be in our atmosphere. What about the harmful stuff that shouldn't and is causing climate change and other mayhem? Air pollutants are harmful (to people and/or the environment) substances of any kind at large in the air. Most of them have been put there as a result of human activity, though some come from natural sources, such as volcanoes or animal farts. Some are gases, while others are particulate matter (tiny bits of solid or liquid suspended in the air). The most significant pollutants are:

Sulfur dioxide and other oxides of sulfur. Produced by burning fossil fuels and other industrial processes (and volcanoes). Responsible for acid rain.

Nitrogen dioxide and other oxides of nitrogen. Produced by burning substances at very high temperatures. A key component of smog.

Carbon dioxide and carbon monoxide. Produced by burning carbon-rich material. Carbon dioxide is an essential atmospheric gas in the right quantities because plants require it, but in excess it works as a greenhouse gas, preventing

the sun's radiation from exiting the atmosphere and thus increasing global warming. Carbon monoxide is extremely poisonous.

Methane and similar hydrocarbon compounds. Hydrocarbon compounds—compounds made of hydrogen and carbon (also called organic compounds)—are poisonous and may act as greenhouse gases.

Particulate matter. This covers a vast array of substances, from asbestos fibers to sea salt and rock dust to carbon. Most of it you don't want to inhale, and it can have dramatically bad environmental effects, too.

Measuring Pollutants

The amount of a pollutant in the air is measured by parts per million by volume (ppmv). Wait, what? What's a part? A million whats? It depends on what you're measuring, but if it's a gaseous pollutant in air, you're going to be talking about milliliters of pollutant per million milliliters of air. With

THE OZONE PARADOX

Ozone is an unstable form of oxygen in which each molecule holds three atoms instead of two: O_3. While its presence in the upper atmosphere is vital to protect us from ultraviolet radiation, at ground level it can be a dangerous pollutant. It is formed by sunlight acting on molecules of other air pollutants, like nitrogen oxides. The ozone layer has been thinning out by about 4 percent each decade since the 1970s, with more dramatic thinning at both of the poles—mainly because of increased chlorofluorocarbon (CFC) compounds, which we use as refrigerants, solvents, and various other things. The CFC molecules lose their chloride and fluoride ions when they reach the upper atmosphere and get battered by ultraviolet rays. These ions then react with the ozone molecules.

particles the measurement is more likely to be micrograms or milligrams per cubic meter of air.

Methods of Measurement

There are several different ways to measure pollutants in the air. **Passive sampling** is the simplest: Someone just unseals one end of a diffusion tube, places it with the open end downward wherever they want to measure air quality, leaves it there for a month, then sends it off to a local laboratory. The tubes contain steel gauze coated with a chemical that reacts with the pollutant the testing body is interested in. Usually diffusion tubes are used to detect nitrogen dioxides, and the chemical used is triethanolamine, which converts the NO_2 to nitrites, which the lab analyzes.

Active sampling involves taking a known quantity of air and testing it for a variety of chemical pollutants. Most air-quality monitoring sites carry out both passive and active monitoring. The active kind is more likely to be used to monitor pollution around specific locations where there's cause for concern at certain times.

Researchers also measure water and soil quality for pollutants, both routinely and in response to natural or human-caused disasters. Drinking water, in particular, is tested for the presence of nasty substances, such as heavy metals, dissolved salts, pesticides, and *E. coli* or other bacteria.

Useful Chemicals from Crude Oil

Crude oil is a fossil fuel, formed from long-dead ancient algae and microscopic water animals which sank down to lake bottoms or sea beds and were then mixed up with mud and squashed under layers upon layers of sediment in an oxygen-free environment. These conditions gradually turned the solid organic material into liquid and gas. A similar process with land plants was responsible for the production of coal.

Crude oil contains lots of different kinds of hydrocarbon molecules. Carbon and hydrogen tend to combine to form rings or long, sometimes branching chains (or combinations of both). Nearly all the vehicle fuels we humans use at the moment are derived from crude oil. Estimates of how long the oil we have left will last varies greatly but most put it at well under a hundred years, even allowing for the reserves that should be out there but that we haven't found yet. Additionally, burning fossil fuels is a major source of pollutants.

Chemicals Distilled from Crude Oil

Distillation is the process of separating out certain components from a liquid by boiling it and condensing back the vapor that is produced at particular temperatures.

Gasoline. Comprised mainly of fairly short-chain hydrocarbons, with between four and twelve carbon atoms per molecule.

Diesel. Gives you more mileage for your money. It contains longer-chain hydrocarbons than gasoline. Jet fuel is chemically quite similar to diesel.

Kerosene. Widely used as a fuel for cooking and heating. Kerosene is also burned with liquid oxygen to make rocket fuel. It is composed mostly of 6- to 16-chain hydrocarbons.

Motor oil. Lubricants for engines of various kinds, to protect moving parts from wear.

Alkenes. Used to make plastics, asphalt, and tar.

LIFE-CYCLE ASSESSMENTS

Assuming we don't annihilate ourselves or our planet anytime soon, our descendents will probably—hopefully—be around to learn about the turn of the millennium as a troubled time for humanity, when the devastating effects of our industries on the Earth were finally undeniable, and we were slowly, reluctantly, and fearfully beginning to find new ways forward.

Taking a realistic and detailed look at the impact of a given product on the Earth, from the manufacture of its components from raw materials to its disposal at the end of its working life, is an important first step to working out how we can reduce that impact. Making the results public means that we consumers can make informed choices about what products and services we use.

The life-cycle assessment (LCA) is a part of the ISO (International Organization for Standardization) 14,000 standards for environmental management in businesses of all kinds.

Ideally, everything gets assessed. In terms of a manufactured product, the extraction of raw materials, the water used in any purification or distillation process, and the fuel burned to transport those materials around all need to be measured. If the product is meat or vegetable, you'll need to assess how much water and other resources it consumes as it grows. The actual manufacturing process itself will use more resources and generate more pollution. Packaging, distribution to outlets for sale, the things the product uses up in its actual function, how recyclable the product is, and the likelihood that it will actually end up being appropriately recycled all need to be assessed. The level

continues

continued

of detail is also important. Do you make a single calculation for road miles covered or factor in relative amounts of congestion where the delivery vehicle can't be driven so economically?

It's not surprising that coming up with simple figures for LCAs is difficult, and two assessments for the same product may produce quite different results. The science involved is in its infancy, and no doubt we'll get better at it. In the first place, though, LCAs should help identify those products that have the most severe environmental impact and point the way to making improvements.

Structure of the Earth

Once we humans figured out that our planet was a sphere rather than a flat disc, we wondered what lay below the visible surface, all the way down to the center of the sphere. No one's been in all the way to check, but the amount of gravity the Earth exerts on other objects in space allows us to calculate its density as a whole, which in turn helps us to work out the composition of the layers that are deeper than we can actually physically investigate. The conclusion is that there are several main layers—solid on top (the crust), becoming viscously flowing semisolid below that (the mantle), then a more fluid liquid layer (the outer core), and finally the very inside (inner core), which is solid.

Elements of Earth

As the Earth formed, its denser materials would have migrated to the center. Iron, one of the most abundant elements on Earth, is thought to constitute about 80 percent of the inner core. It has been proposed by some scientists that the core is actually a single colossal crystal of pure iron, but most believe it is mainly iron with some nickel mixed in—nickel is a close neighbor to iron in the periodic table. The extreme pressure on the inner core means the iron stays solid at temperatures that would normally melt it.

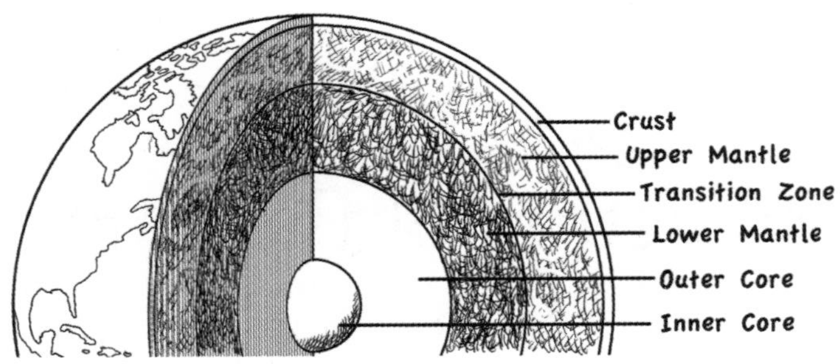

The outer core is under less pressure, but it, too, is screamingly hot, so the metal of which it is made (again, probably a mixture of mainly iron with a bit of nickel) is liquid. All this iron in the outer core is responsible for the magnetic field of Earth.

The mantle sometimes makes an appearance via volcanoes. It is composed of molten rock, which is rich in silicon, magnesium, and, once again, iron, as well as tiny amounts of a whole range of other elements. The iron and most other metals here are not pure, but held in oxides and other compounds. There is a continuous exchange going on between the molten rock of the mantle and the solid rock of the crust. Volcanoes release **lava**—molten rock from the mantle—which in due course cools down and becomes surface rock. This gets eroded by wind and water, and its dust is deposited elsewhere as sediment. The buildup of sediment means that the lower layers are compressed down over immensely long time spans until the heat and pressure melts them back into the mantle. *Et voilà,* the rock cycle.

The eight most common elements in the Earth's crust by mass are oxygen (46.6 percent), silicon (27.7 percent), aluminum (8.1 percent), iron (5 percent), calcium (3.6 percent), sodium (2.8 percent), potassium (2.6 percent), and magnesium (2.1 percent). Another 84 elements make up the remaining 1.5 percent or so.

BURIED TREASURE

Why do we like gold, silver, and platinum so much? One reason is that when we find them in their natural state, they are recognizably pure gold, silver, and platinum rather than grubby-looking oxides. These metals are very nonreactive, which means that they don't react with everyday elements like oxygen but instead stay lovely and shiny and are ideal for decorating stuff, including our fingers, necks, and earlobes. All three are sometimes described as "noble metals," drawing comparison with the nonreactive noble gases. Other noble metals include rhodium, palladium, iridium, and, according to some, mercury, but they are not so pretty and shiny—and, in the case of mercury, since it is liquid at room temperature, rather messy to wear. They are all very rare in the Earth's crust, making them even more valuable to us.

Types of Metals

Metals have the special property of metallic bonds, which gives them many of the qualities that make them stand out from the other elements. As well as being flexible and shiny, they are also excellent conductors of electricity and heat. These attributes make metals very important and useful to us.

Most of the elements in the periodic table are metals. You'll find those that aren't metallic clustered away on the right-hand side of the table, and between the two is a diagonal dividing line of elements like silicon, which have some metallic traits and are known as metalloids. Groups 14 to 16 in the table all transition from nonmetal to metal as you go down, via two metalloid elements.

Base metals refer to a variety of metals that corrode (rust) quite readily. The so-called **noble metals** (as discussed in the previous section) don't. Scientists have also synthesized some compounds that have metallic traits, including a silvery-gray sheen and ability to conduct electricity.

Alloys

If you look around your immediate surroundings, you'll probably quickly find a variety of things that are clearly made of metal. However, it's a fair bet that none of them are made of a single, pure, elemental metal. Often, mixing two metals or a metal and a nonmetal together to form an alloy produces a more useful product. Pure gold, for example, is very soft and easily knocked out of shape, so gold used in jewelry or coins is typically an alloy of gold and one of the base metals, to give it a bit of extra hardness. Bronze is an alloy of copper and tin, while brass combines copper and zinc.

Steel is a very widely used alloy of iron and carbon; adding a small percentage (up to 2 percent) of carbon significantly strengthens the material. However, throw in too much carbon and the material becomes brittle. Adding chromium to steel produces stainless steel, which does not rust.

An alloy is not a compound, because the components don't react together, but it is a solution, because the new element's atoms are completely mingled within the metallic matrix. In steel the carbon atoms fit between the iron atoms without disturbing the repeated-cube structure assumed by the iron atoms. When the structure is placed under stress, the carbon atoms help "fill the gaps" when cracks form in the metal matrix.

Construction Materials

Our ancestors established the use of various metals for making all manner of tools. Alloys of various kinds are suitable for different types of construction, and using metals alongside nonmetallic materials, like stone and wood, with their range of different properties, has enabled us to undertake construction on a massive scale, dramatically changing both everyday human life and the whole topography of the planet.

When you're building a large structure—say, a bridge or a house—you are going to want it to stand for a long time, coping with a whole range of everyday forces. A bridge across a river must be strong enough to bear the weight of constant traffic, and it must be resistant to corrosion and erosion from moving water. The house must be strong enough to support its floors and the furniture that stands on them, and it must keep the weather out and the warmth inside. Measurable properties of construction materials include compressive strength (resistance to being squashed), tensile strength (resistance to being stretched), density, porosity, conductivity of heat and electricity, fire resistance, and sound insulation.

Stone

Buildings intended for people to live in are usually made out of some kind of rock or stone because most are strong, durable, and have good insulation properties. The basic rock raw product is usually cut or formed into bricks or blocks

WHAT KIND OF METAL?

Steel is the most popular metal in construction today. The proportion of carbon to steel determines its strength. We have already seen how mixing steel with chromium produces stainless steel, which resists corrosion more effectively and is also attractively shiny. It is mostly used for small tools, such as cutlery and surgical instruments. Steel used in construction also needs to be as rustproof as possible, and a different method is used to achieve this: galvanization. Steel is galvanized by putting it in an 860°F (460°C) zinc bath. In contact with air, the zinc reacts with oxygen, and then the resultant zinc oxide reacts with carbon dioxide. The result is a coating of zinc carbonate, a dull inert compound that prevents further reaction. Galvanized steel is a tough, strong, and very durable material, suitable for all kinds of outdoor constructions.

---

that are easily stackable and often fixed together with cement or another bonding substance. Most modern bricks and building blocks are made from fired (heat-dried) clay or set concrete rather than being cut from solid stone. This means the manufacturer can control the consistency of the material as well as the shape of the blocks and incorporate air spaces, if necessary.

Metals in Construction

Metals in small amounts are found in pretty much every kind of large-scale construction. The right alloy can offer enough hardness and resilience to stand up to extremely pressured conditions. However, some of the other metallic properties, such as its high heat conduction, mean it's not suitable for everything.

Metal is a good choice for long bridges because of its high strength-to-weight ratio and also its high tensile strength, a property lacking in concrete and other stone-based materials. For similar reasons, metal is used for roofing, especially on large buildings. It is easy to coat metals with reflective paints, which can have significant cooling effects when used to roof buildings in hot environments.

Organic Chemistry

Organic chemistry involves the study of the structure, properties, composition, reactions, and preparation of carbon-based compounds—and that includes all life forms, which are carbon-based, at least on planet Earth. We know of or have constructed more than 6 million organic compounds, including all life forms, the foods we eat, plastics, synthetic and natural fibers, dyes and drugs, insecticides and herbicides, ingredients in perfumes, flavoring agents, and petroleum products.

Natural Polymers and Their Roles in Nature

A **polymer** is a large molecule with a repeated structure, forming a chain. The single "links" in the chain are called **monomers;** join up lots of monomers and you produce a polymer. Here's an example: The monomer molecule ethylene can join up to form the polymer molecule polyethylene when the double bond between ethylene's two carbon atoms is broken, leaving each free to form a new bond with another carbon atom.

```
H         H                                 H   H   H   H   H   H   H   H
 \       /                                  |   |   |   |   |   |   |   |
  C  =  C       ------------------->   ---C - C - C - C - C - C - C - C---
 /       \        Polymerization            |   |   |   |   |   |   |   |
H         H                                 H   H   H   H   H   H   H   H
   Ethylene                                            Polyethylene
```

Most natural polymers look a lot like this. They are based around a long chain of connected carbon atoms, with hydrogen atoms coming off to the side. Hydrogen and carbon-based polymers are called organic because of their prevalence and importance in living things, although we can also synthesize them artificially. If they contain just hydrogen and carbon, they are **hydrocarbons**—we looked at lots of them in the section on crude oil. If they incorporate oxygen also, they are **carbohydrates.** Have you noticed by now that molecules incorporating oxygen often end with "ate." Polymers may also include atoms of many other elements.

What Are Polymers For?

Lots of things in nature are made of polymers. One of the most important is DNA, or deoxyribonucleic acid, the molecule that provides the "instructions" for making all the different kinds of proteins in our bodies. DNA is a pretty complex polymer; actually, to be precise, it is two pretty complex polymers because each strand of DNA comprises two polymer chains connected by a "ladder" of hydrogen bonds and twisted into the familiar double-helix spiral. The monomers are **nucleotides,** each made of a simple sugar, a phosphate group, and one of four kinds of nucleobase—don't worry too much about what all this means. For now the point is that the pattern is repeated, again and again, to form a chain of (theoretically) endless length.

So DNA is one kind of natural polymer. Another one, simpler but equally useful (if you happen to be a plant), is

cellulose, which is the main constituent of plant cell walls. It is also the main source of dietary fiber, which helps to protect us from unpleasant intestinal issues, so it's actually not just plants that consider it important. Cellulose is a carbohydrate, composed of carbon, hydrogen, and oxygen, and its structure is expressed like this:

$$(C_6H_{10}O_5)n$$

The "n" shows that the structure is repeated as many times as you like. As a diagram, it looks like this:

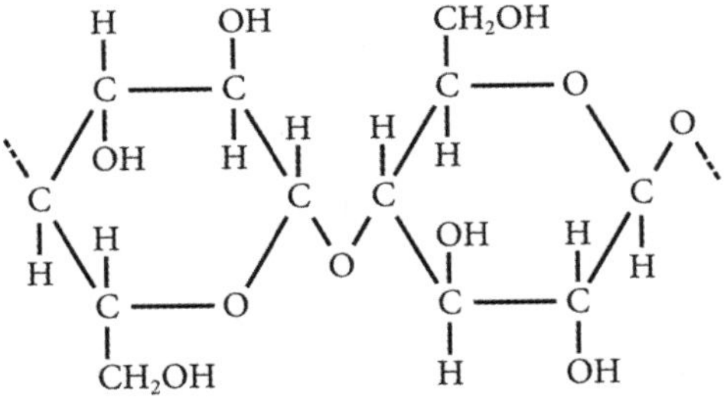

```
        H      OH          CH2OH
        |      |             |
        C------C             C------O
       /|      |\ H     H  / |       \    O
  \   / OH     H \|     | /  H        \  / \
    C             C     C              C
    |\ H         / \   / \ OH      H  /|
    H \|        /   O   \|         | / H
       C------O          C---------C
       |                 |         |
     CH2OH               H         OH
```

The ring-shaped monomers that make up carbohydrate molecules like cellulose are simple sugars. Therefore, molecules like these are sometimes called **polysaccharides**—"many sugars." Another name for polysaccharides is starches. Our bodies break down starches, like glycogen—$(C_6H_{10}O_5)$ *n*—to release simple sugars like glucose, which are then used in chemical reactions in our cells to provide energy.

POLYUNSATURATED?

Another kind of polymer you'll find in your cells is fatty acid. These molecules have lots of jobs within our bodies, and we eat them in the form of fats. You probably know that polyunsaturated = good and saturated = bad, but what's the difference? A saturated fatty acid is one in which all the carbon atoms (except the one at the end of the chain—the "acid" part) are joined to two hydrogen atoms, and to each other by single bonds. If one or more adjacent pairs of carbon atoms have only one hydrogen atom each, they are joined to each other by a double bond instead. Because there is "room" for additional hydrogen atoms in the molecule, it is "unsaturated"—monounsaturated if there is one double bond, polyunsaturated if there are several. No room for more hydrogen = saturated.

The essential fatty acids we all need to consume because our bodies can't make them are unsaturated. However, we can make

Saturated fatty acid

Unsaturated fatty acid

our own saturated fats, so we really don't need to eat them. They get stored as a long-term energy source, and too much of them makes us fat.

Trans fats are unsaturated fats in which the carbon chains are arranged in straight lines rather than with kinks at each double bond. They don't occur widely in nature—most of the ones we eat are formed from processed vegetable oils used in delicious fried junk food. The innocent-sounding structural difference makes trans fats extremely bad for us because they are much harder for the body to break down and digest. Such is the wonder of chemistry.

Nutrition

We're made of chemicals, and we use up chemicals constantly in our innumerable bodily processes, so we need to eat chemicals in order to keep living. The most convenient way is to eat the bodies of other living (well, living until recently) things, which are made up of the same kinds of chemicals as us. Dietary chemicals are called **nutrients**.

A Balanced Diet

We need to eat proteins, carbohydrates, and fats. Carbohydrates and fats are both made up of carbon, hydrogen, and oxygen. Proteins, which contain nitrogen as well as carbon, hydrogen, and oxygen, are used to build structural tissues and also form "working" molecules like antibodies, enzymes, and some hormones.

The Essentials

Some of the chemicals we need in our body are things we can synthesize ourselves. Others we cannot, and we have to eat them instead. Of the 20 "standard" amino acids we use to construct our proteins, 8 are called "essential." They are essential because we have to include them in our diets. They are isoleucine, leucine, lysine, methionine, phenylalanine, threonine, tryptophan, and valine. The disease kwashiorkor is caused by a deficiency of essential amino acids. Luckily for most of us, the essential proteins are all found in commonly consumed foods like most kinds of meat, grains, beans, and eggs.

There are essential fatty acids, too. You may be among the many who takes them in the form of supplements—they are not as widely available in "normal" foods as the essential amino acids. They come in two kinds: omega-3 (alpha-linolenic acid) and omega-6 (gamma-linolenic acid). What about omega-9? I hear you ask. It exists, but it's not normally essential to include in our diet, since most of us can manufacture it in our bodies. Those with compromised liver function sometimes can't, however, which is why supplements often contain all three kinds. To avoid the need for supplements, make sure your diet includes fish and/or various oily seeds, including those of flax, sunflower, and pumpkin.

Vitamins, Minerals, and Fiber

Vitamins are specific compounds that are vital (albeit often in tiny amounts) for healthy life and which we can't manufacture ourselves (or make in sufficient quantities), so we have to eat them. Even quite closely related animals may vary in terms of which vitamins they need. For example, most mammals, including primates, can make their own vitamin C, but we humans can't, and nor can guinea pigs. (Guinea pigs, however, seem to have a much greater willingness to eat their greens than we do.) Vitamins are mostly pretty simple molecules. Some are carbohydrates, some proteins. But they are not related in terms of their chemical structure.

Dietary minerals are how we get hold of the other elements our bodies need, besides carbon, hydrogen, and oxygen. For example, our red blood cells contain iron; some enzymes need zinc, copper, and selenium; our bones contain magnesium;

and phosphorus, sodium, and chlorine have multiple uses. Because these elements are necessary for other animals and plants, too, we usually get enough of them in our regular diets, but taking supplements is popular, too.

Dietary fiber is made of the cellulose that forms the tough cell walls in plants. Even though it mostly passes through our systems undigested, its passage is necessary to keep our digestive systems—especially our colons—in good working order. Some of the ways in which we process foods—turning whole-wheat grains into white flour, for example, amounts to removing the cellulose. In many Western cultures people don't eat enough fiber and have high rates of bowel trouble.

Harmful Chemicals

It seems like a paradox, but sometimes the chemicals that do real damage to our bodies are structurally similar to those that are actually supposed to be there. They disrupt normal body processes, usually by chemical reactions that deactivate, destabilize, or otherwise interfere with the body's own molecules. Some substances, which are beneficial and utilized by the body in smaller amounts, can be poisonous in larger quantities.

Poisoning may be a rapid and dramatic process or very gradual and cumulative; it can be severe or mild, depending on the type of poison or toxin and the amount consumed. Poisons may target the nervous system, destroy blood cells, and keep the liver from working, among other unpleasant things. Many poisonous substances can be cleared out by the body over time—alcohol is a good example. But often there is an amount beyond which there's no return—the lethal dose. Some poisons are harmless when ingested, but bodily processes turn them into toxins. Many of the medicines we use are poisons, designed to inhibit a natural process that is happening to excess.

Poisons in the Diet

Some living things manufacture poisons in their bodies to protect themselves from being eaten. If a bird eats a little fluffy white moth and promptly feels ill because of toxins in the moth's body, it is likely to avoid fluffy white moths in the future, so as a whole, that species of fluffy white moth benefits. We know (and presumably our ancestors learned the hard way) that not every shiny red berry is delicious and nutritious. Our accumulated knowledge of what is and isn't safe to eat has resulted in people rarely poisoning themselves this way today.

Pesticides are tailor-made to be poisonous to kill plants and animals that interfere with the crops we grow, so it's no surprise that they can poison us, too. After World War II the chemical DDT, or dichlorodiphenyltrichloroethane, was widely used as an insecticide. By the 1960s severe declines in many bird species were linked to its use. The chemical accumulates in animal bodies, meaning that birds that eat poisoned insects carry a DDT "load" that is then passed on to the next species in the food chain. The effects on birds of prey were particularly devastating. Its use has been banned for decades now, but it is an extremely persistent chemical, and its residues can still be found in the fat stores of livestock and fish.

Airborne Poisons

Air pollution gets into our systems via our lungs. The solid particles don't get through into our bloodstreams, so they stay

in our lungs and slowly clog them up. The gaseous molecules do get into our blood, and from there they can cause havoc. One of the most dangerous is the odorless and colorless carbon monoxide. At least with sulfur dioxide you know it's there by its horrible smell, and nitrogen dioxide is similarly smelly and has a red-brown color. Carbon monoxide poisons by reacting with the hemoglobin in red blood cells, making them unable to carry oxygen.

VENOM

In biology, poison that an animal administers to another via fangs or a sting is called venom, and the animal that does it is venomous. The most venomous animal on Earth? The golden poison dart frog in South America.

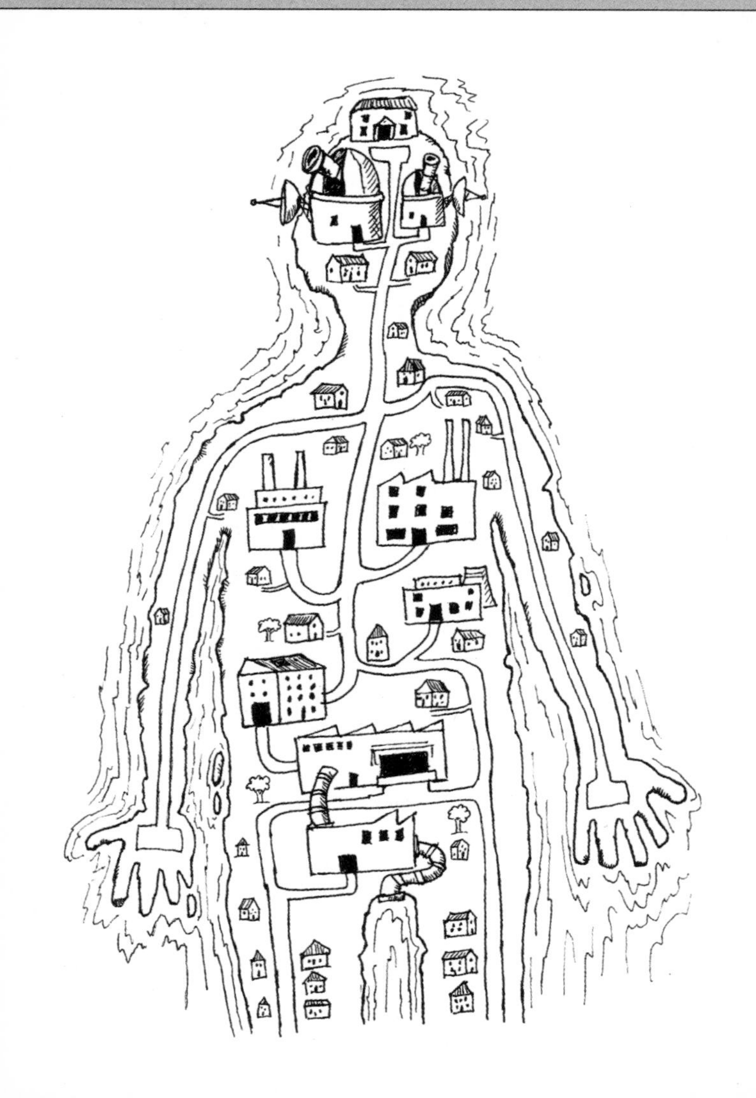

3

Biology

Human (and Other) Bodies

A body is rather like a city. The organs are like buildings, some used for storage, some for production, some for processing. The "workers" and the "products"—blood cells, hormones, molecules of food, and so on—are like vehicles traveling from organ to organ as required.

Circulation

Our internal transport network is our circulatory system. The most important part of this network is the part that carries our blood: the cardiovascular system.

Oxygen and Blood Vessels

The oxygen we breathe in is needed by most of the cells in our bodies to carry out essential functions. One of the most important "jobs" of blood is taking this oxygen to where it's needed, then bringing the blood back to the lungs to load up again with fresh oxygen.

The main blood vessels that carry oxygenated blood away from the heart to the body are **arteries.** They divide up into

BLOOD COMPOSITION

There are three main constituents of our blood:

Red Blood Cells. Properly called **erythrocytes,** these are small cells with a squashed disc shape. They contain hemoglobin, which combines with oxygen, and unsurprisingly, they give blood its red color.

White Blood Cells. Also called **lymphocytes,** these are part of our immune system. Their role is to deal with any bacteria that enter our bloodstream. White blood cells include **platelets,** which are involved with blood clotting when a blood vessel is broken.

Plasma. A pale yellow liquid that's mostly water, plasma is there to carry the cells around. Hormones and food molecules also travel "loose" in the plasma.

---

smaller vessels called **arterioles.** These divide again into tiny vessels: **capillaries.** Oxygen exchange happens through the walls of the capillaries. Unlike the larger blood vessels, capillary walls are just one cell thick, and through them oxygen and other substances can be exchanged.

The deoxygenated blood proceeds from capillaries to **venules** (equivalent to arterioles), which in turn join the large **veins** (equivalent to arteries) that eventually return the blood to the heart.

Arteries are thicker than veins, and their walls contract to help keep the blood moving, producing the pulses we can feel in our necks, wrists, and other places. Veins are less muscular, but the squeezing of our surrounding muscles helps keep blood moving through them (that's why brides and grooms faint at the altar if they stand still too long), and valves prevent blood "backflow."

The Heart

The heart is, basically, a blood pump. As organs go, it's rather simple; it is divided vertically into two halves, each with top and bottom chambers. The left side of the heart receives freshly oxygenated blood from the lungs into the top chamber—the left **atrium,** via the pulmonary vein. The blood proceeds into the lower chamber, the left **ventricle,** which pumps powerfully to send that blood off to the rest of the body via our largest artery: the **aorta.** The right atrium receives the deoxygenated

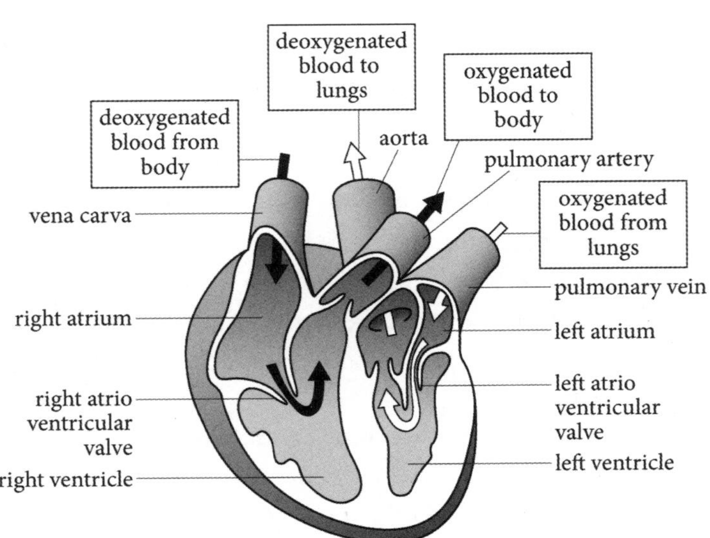

ARTERIES AND OXYGEN

Arteries go out from the heart, and veins return to it, so in nearly all cases, you'll find bright red oxygen-rich blood in arteries and darker, oxygen-depleted blood in veins. The exception is with the pulmonary artery and vein, which travel to and from the lungs respectively.

---

blood via our largest vein, the **vena cava.** It then travels to the right ventricle, which pumps it off to the lungs via the pulmonary artery to collect more oxygen.

The walls of the heart are made of cardiac muscle, which works constantly and unconsciously (good thing, too). Ventricles are thicker-walled than the atria, because they do more serious pumping. Valves between atrium and ventricle and ventricle and aorta/pulmonary artery regulate the blood flow. It is the closing of these valves that produces our heartbeat.

Lymphatic System

Alongside our cardiovascular system is the lymphatic system, which is also made up of a network of tubular vessels (but it lacks a central pump). It carries lymph—a clear fluid similar to blood plasma. The lymphatic system is a go-between for blood and body cells, moving around such things as newly formed white blood cells, food molecules, and excess interstitial fluid (the fluid that surrounds all of our body's cells).

Circulation in Other Organisms

Because the heart of an insect carries only food, not oxygen, its blood is green. The heart itself is a simple muscular tube. Other vertebrates have simpler hearts than birds or mammals, with fish having just a single atrium and ventricle. Plants transport water around their various parts via a network of **xylem** vessels, while **phloem** vessels transport food in the form of sap.

Skeletal Structure

Bones have four main functions: They are anchor points for the muscles we use to move around, thus providing support and structure to our otherwise rather squishy bodies; they protect our particularly squishy and important organs from damage, and they store minerals. The bone marrow is the factory that manufactures our blood cells.

The Human Skeleton

As we all know, the head bone is connected to the neck bone, and so on—except that it's much more complex than that. That old song would need to mention 206 individual bones in adults (babies have more, but some fuse as they grow) to be strictly anatomically accurate. We'll aim for the middle ground and describe the main groups of bones.

Skull. The skull protects the brain. The lower jaw or mandible is a separate part. The tiny ossicles are three bones in each inner ear that help us to hear.

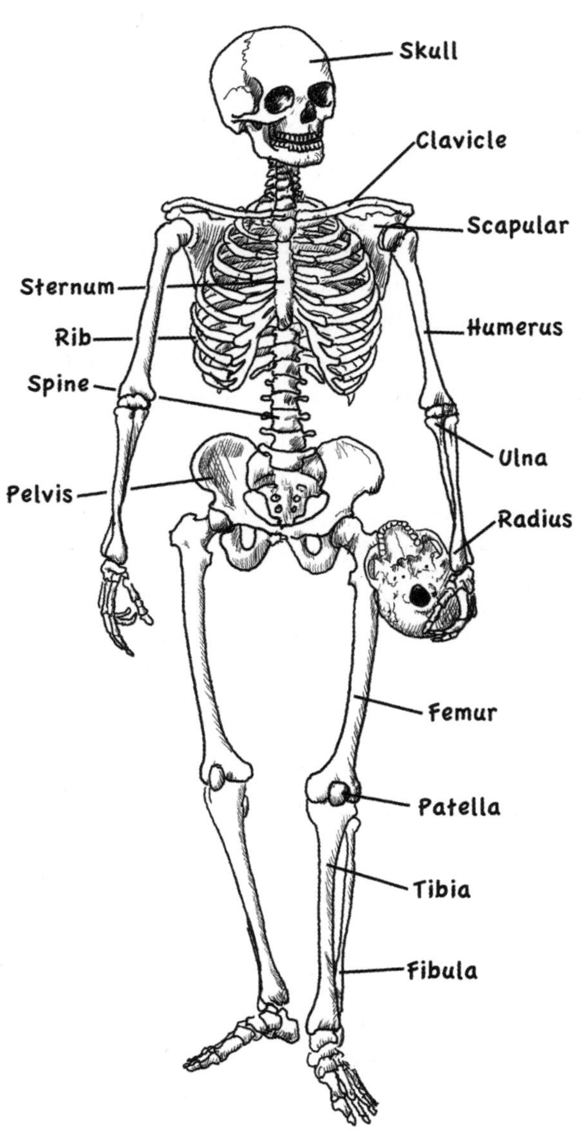
Skull
Clavicle
Scapular
Sternum
Humerus
Rib
Spine
Ulna
Pelvis
Radius
Femur
Patella
Tibia
Fibula

Spine. The vertebral column, or spine, is comprised of 33 vertebrae, which are chunky little bones with holes in their centers through which the spinal cord runs. The column is traditionally divided into four sections—cervical, thoracic, lumbar, and pelvic (or neck, upper back, lower back, and bottom).

Limb Bones. We have one long bone in the upper limb (**humerus** in the arm, **femur** in the leg), two long bones in the lower limb (**ulna** and **radius** in the arm, **tibia** and **fibula** in the leg, with the kneecap or **patella** between them, and the femur). **Carpals** and **metacarpals** are in the wrist and hand, **tarsals** and **metatarsals are** in the ankle and foot, and **phalanges** are in the fingers and toes.

Trunk Bones. Finally, the bones in the middle consist of the **rib cage,** which protects the main organs, with the **sternum** or breastbone uniting the upper ribs at the front of the body. The **scapulars** (shoulder blades) and **clavicles** (collarbones) connect the arms to the upper body, and the **pelvis** connects the legs to the lower body and also helps to protect lower-body organs.

HANGING IT ALL TOGETHER

Bones connect to each other via joints, which come in different kinds and allow for different degrees of movement. **Synovial joints,** like the ball-and-socket joint of the shoulder or the hinge joint of the knee, allow very free movement. These complex joints are protected by pads of cartilage and lubricated by synovial fluid. Other bones are joined by bands of **cartilage** or fibrous tissue and allow minimal movement.

You'll find essentially the same bone groups in most other vertebrates, though they vary a lot in structure and relative size. For example, the inner ear bones in mammals are found in the jaws of reptiles, where they have a completely different function. Invertebrates have rigid outer coverings (**exoskeletons**) instead of internal skeletons or do without "hard parts" in their bodies altogether.

Bone Building

Crack a bone in half and you'll see that the outer layers are hard and mineralized, but inside it is soft and spongy. The outer layers are made of inactive bone-forming cells **(osteoblasts).** These layers are dense and compacted; they contain lots of calcium and phosphorus. The spongy bone inside contains active bone-forming cells, which mainly make the tough protein known as **collagen.** It has a generous blood supply and includes the red, soft bone marrow that manufactures blood cells of all kinds.

Muscles and Skin

If you want to turn heads on a nudist beach, you'll want your muscles and skin to look their best. Appearances aside, the two form distinct systems. Your muscles move you around. Your skin is your interface with the world outside—it provides a physical barrier but also enables you to take in some of the things your body needs and to get rid of some of the things it doesn't.

How We Move

Two of the three kinds of muscle—**cardiac muscle** in the heart and **smooth muscle** in places like our digestive system—are **involuntary** and carry on doing their thing as long as we're alive. **Skeletal muscle,** however, is **voluntary** muscle—we (usually) decide when we want it to do something, although often on a subconscious level.

When a muscle is contracted, it shortens. In the case of the biceps muscle, which is on the top of the upper arm and is connected at the shoulder and the top of the forearm, its contraction causes your arm to bend. Conversely, the triceps, which is on the back of the upper arm, causes your arm to straighten when it is contracted. The abdominal muscles cause your middle to curl forward when contracted—muscles in your lower back straighten you out again. Most body movements involve several muscles at the same time, working to a greater or lesser degree (just as those diagrams on the weight machines at the gym show you). Some other muscles provide stability rather than movement when contracted.

The human body contains more than 600 skeletal muscles. All of them are made of bundles of muscle fibers enclosed in a tough membrane sheath. At their ends they become tendons, which attach to bones. The contraction within the muscle fibers is caused by bonds forming and breaking between two protein types, **actin** and **myosin,** in response to electric signals from motor nerves (the ones that send messages from the brain to the muscles when we want to move).

SKIN COLOR

The amount of **melanin** in our skin gives us our skin color. Darker skin is more common in hot places because the melanin acts as a sunscreen, protecting against dangerous ultraviolet radiation. However, for the same reason, high levels of melanin make us less able to make vitamin D, which is manufactured in the skin in response to sunlight. So in less sunny places, people tend to have less melanin. Dark-skinned people who live in northern places with less sunshine need to make sure they get sufficient vitamin D in their diets, just as pale-skinned people need to take extra care to protect their skin with sunscreen in hot countries.

Outer Covering

The largest human organ is the skin. It is a flexible continuous covering that is over nearly all of the body. It keeps bacteria out, regulates our temperature and water content, helps us make vitamin D, and provides sensory information about the world around us.

The outer layer of skin—the **epidermis**—is composed of dead epithelial cells, which are continuously worn away and replaced. Hairs grow out of it, and pores within it are outlets for sweat glands. At the bottom of the epidermis is the pigment layer where the skin pigment melanin is found. Farther down is the **dermis** layer, wherein are the follicles of the hairs, with their associated erector muscles and sebaceous glands, the sweat glands themselves, blood vessels, and sensory nerve endings of various kinds. Under these is a layer of fat—the **subcutis**—which provides insulation.

Sebaceous glands produce an oily substance called sebum, which helps keep skin and hair soft and flexible. **Sweat glands** release salty water, which helps to cool us down by evaporation and regulates our salt and water levels. The blood vessels close to our skin can widen when we need to lose some heat—that's why we get flushed after exercise—and they can constrict to retain extra heat when we're cold. **Sensory receptors** in our skin detect pressure, vibration, heat, cold, and pain. We can absorb some substances through our skin (sadly, not always things we would want to).

FUR COATS

Each of a mammal's body hairs has a little muscle that can make the hair stand on end. While this is pretty useless in humans, in furry critters it can make a huge difference for retaining and losing heat. Fur standing on end also makes the critter look larger, which can help it scare off a rival or potential predator. Fur is also a handy "canvas" for colors, patterns, and markings.

Nervous System

Boo! When startling things like that happen, you become very aware of your nervous system. But it's also busily working away in the background all the time, because it controls a vast array of conscious and unconscious bodily processes—in fact, all of them.

Brain + spinal cord = the central nervous system. The outbranchings of the spinal cord form the peripheral nervous system.

The Brain

The brain is the center of the nervous system. It is basically a big mass of **neurons,** up to 33 billion of them, along with a larger quantity of **glial cells** that have various roles but basically take care of the neurons. All of this is organized into structures and regions with distinctly different functions.

Cerebral Cortex. These wrinkled, convoluted domes include the parts concerned with perception (interpreting sensory information) and conscious thought. They are proportionately much larger in humans than in other animals.

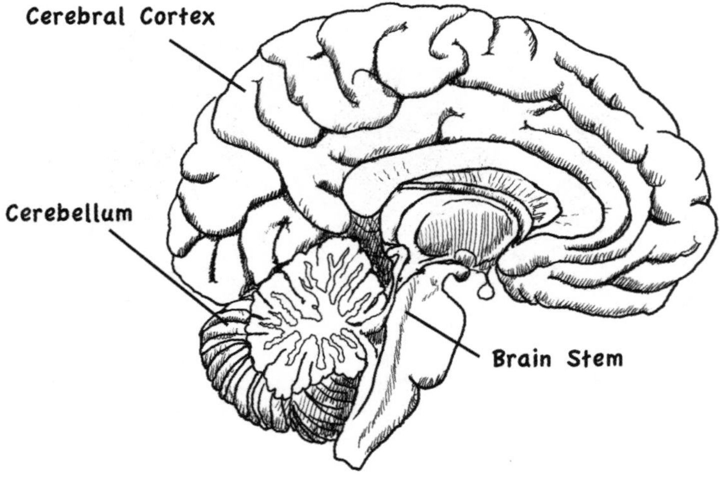

A SINGULAR CELL

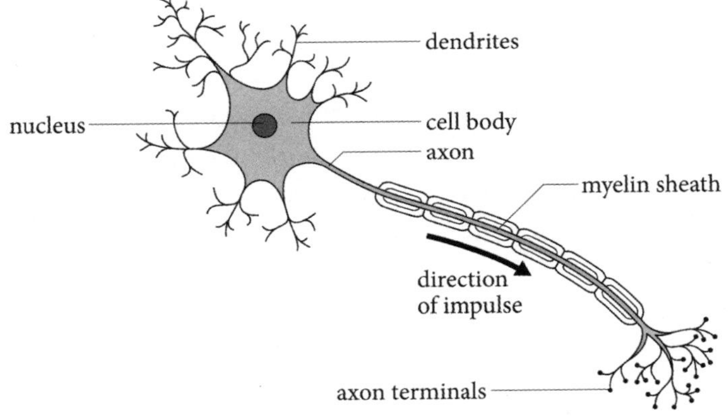

The nerve cell, or **neuron,** is quite a departure from the generic blob-shaped cell with the central nucleus, at least shape-wise. At one end is the cell body with a nucleus—it has numerous long-branched projections called **dendrites.** Another very long projection forms the **axon,** down which the electrical charge of a nerve impulse travels. The axon is insulated with wads of fatty **myelin,** enabling the nerve impulse to travel faster as it jumps from gap to gap between the wads. The other end of the cell is formed by the "foot" of the axon, where there are several bulb-tipped terminal branches. These release a chemical (**neurotransmitter**) that activates receptors on the next cell, so the impulse continues to its destination. Nerves are bunches of axons gathered together like cables.

Cerebellum. This is the tightly wrinkled round bit at the back under the cerebral cortex. The cerebellum is mainly concerned with automatic physical movement (things most of us can do without thinking about it, like standing up, walking, and running). Other brain structures nestle deep inside and deal with a wide variety of other unconscious functions.

Brain Stem. Emerges from the base of the brain and becomes the spinal cord, a thick bundle of nerves that travels down the center of the vertebral column, sending off branches to the various body parts as it goes.

Somatic and Autonomic Nervous System

The peripheral nervous system has two components. There is the somatic nervous system, which deals with voluntary body movements and sensory inputs, and there is the autonomic nervous system, which handles mostly unconscious stuff, like heartbeat, digestive processes, and so on. The two systems use different nerve pathways.

Digestive System

Our bodies are made up mainly of proteins, fats, carbohydrates, and water. We are constantly using up and breaking down these substances in physical processes, such as growth, replacing broken parts, and moving around. So we need to keep on putting more in, and we do this by eating plants and other animals—our digestive tracts do the rest.

FIGHT OR FLIGHT?

Back to that startling "Boo!" What happens to your body when you are frightened? The autonomic nervous system kicks in—specifically, the part of it known as the sympathetic nervous system. It brings about a whole range of physiological changes designed to get you ready to deal with the danger or to run away very fast. Adrenaline is released to get your heart rate up. Your pupils widen to better see what's happening, while your digestive tract slows down, so blood is freed up for the muscles, where it's needed. The counterpart system, the parasympathetic nervous system, basically does the opposite and calms everything down again.

Making Tracts

The journey from one end of the human digestive tract to the other is a circuitous one, covering about 21.3 feet (6.5 m) in the average adult man. The journey time is, if anything, even more surprising, normally taking 35 hours or more. All that time and distance is necessary to convert your supper into a collection of molecules small enough to be absorbed into the bloodstream.

Swallowed food goes down your **esophagus** to your **stomach,** which vigorously churns it around and steeps it in strong acids for 3 to 5 hours of processing, after which it has all moved on to the **small intestine.** Travel through these takes about 3 hours, and then it's on to the **colon,** where transit

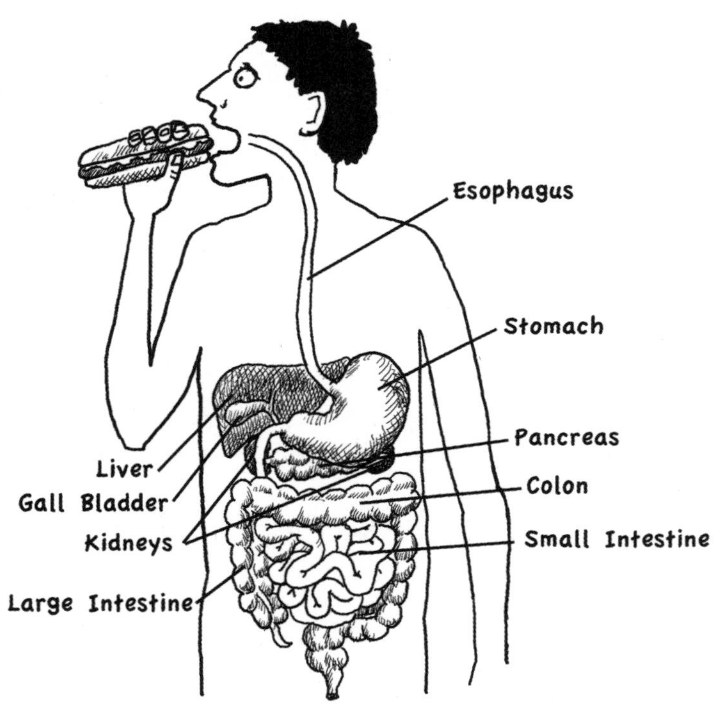

takes up to 40 hours (but can be a lot less). The **pancreas, kidneys, liver,** and **gall bladder** all play a role in the process but are not part of the tract.

Soaking It Up

The main role of the small intestine is food absorption. The inner walls of this long and fantastically twisty tube are lined with fingerlike projections called **villi,** through which food molecules can pass straight into blood capillaries. Most of

ENZYMES

Along the digestive tract, enzymes are released to help break down the food. Enzymes are proteins, and they break down food into successively smaller molecules. Different enzymes work on different nutrients—proteases break proteins down to amino acids, amylases turn complex carbohydrates into simple sugars, and lipases break up fats into fatty acids. The first enzymes are produced by the tongue, and then the pancreas releases more into the food as it leaves the stomach. Acids are produced in the stomach, and later more are formed in the liver and released by the gall bladder, helping enzymes access their target food molecules.

---

the breakdown of food by pancreatic enzymes takes place in the 10-inch (25-cm) **duodenum,** the first short stretch of the small intestine, while the 6.5-foot (2-m) **jejunum** does most of the nutrient absorption, and the 11.5-foot (3.5-m) **ileum** does the rest.

Waste Disposal

By the time the food reaches the **colon,** practically all of the useable nutrients are gone. The colon's main role is to absorb excess water from what's left, though it also takes in a few vitamins that the small intestine doesn't. Bacteria that live in the colon break down dietary fiber here, and quite a lot of them end up becoming part of the final waste product themselves.

When it comes to urinating, excess water in the blood is filtered out by the **kidneys** and leaves the body via the **urethra** after a decent quantity of it has built up in the **bladder.**

Other Animals

Birds may be toothless, but to compensate they have a couple of extra organs in their digestive tracts. The **crop,** an enlarged, stomachlike section of the esophagus, stores food prior to digestion; its contents may be regurgitated to feed the baby birds. The **gizzard** is a muscular grinding "second stomach"—your budgie eats grit to provide its gizzard with extra seed-crushing power.

The stomachs of cows and other ruminant animals are divided into four chambers. Food swallowed into the first (the **rumen**) may be brought back up again for some re-chewing before being swallowed again and moved on to the other chambers for digestion.

Reproductive System

Here's the part you probably turned to first in your school biology textbook, to check for naughty illustrations. Like most animals, we are very highly motivated to have sex—passing on our genes is as much a part of the game of survival as is finding enough to eat, drink, and breathe, though usually considerably more difficult to accomplish.

Male Parts

The male reproductive system is simpler than the female's, so let's begin there. There are two **testes** (testes is the correct plural of testicle) that hang outside the body in the sacklike **scrotum.** Their function is the manufacture of sperm cells, and they require lower temperatures than are found within the body to best accomplish this. A narrow tube, called the **vas deferens,** carries sperm to the **penis** when a man ejaculates. On the way, the prostate gland secretes various fluids, which are added to the sperm to produce semen.

Female Parts

Most of the female reproductive apparatus is internal. Of the external parts, the **clitoris** is a small and highly sensitive organ, analogous to the penis but with the sole function of making sex fun. The **ovaries,** which are analogous to the testes, are located inside the lower abdomen. They contain the eggs, the female sex cells. Females don't constantly make new eggs; at birth they already have all that they'll ever produce. Once a month a single (usually) egg matures in one of the ovaries and is released into the **Fallopian tube.** The two Fallopian tubes connect to the top of the **uterus**, or womb, which becomes lined with thick, blood-rich, spongy material in the days leading up to the arrival of the egg. If the egg isn't fertilized, it is released through the **vagina,** along with the uterine lining; this is menstruation. If the egg has been fertilized, it implants in the uterine lining and begins to grow.

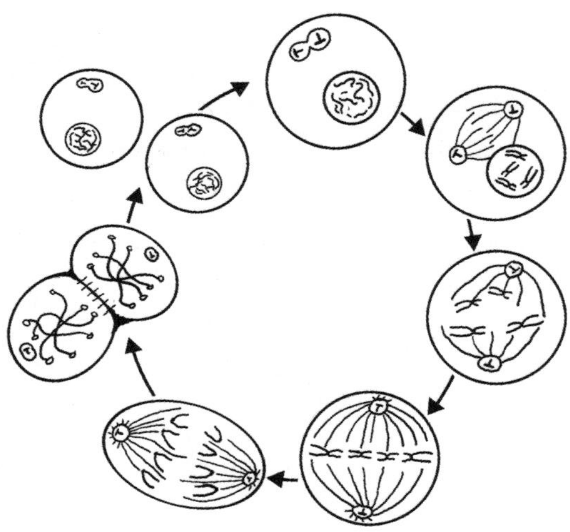

Splitting Cells

Normal cell division is called **mitosis**. It begins with the duplication of all the chromosomes (sets of genetic material) in the cell nucleus and results in the formation of two identical cells from the single parent cell (see illustration above). This process happens constantly in many different parts of our bodies. Things are different with sex cells. Their division is called **meiosis,** and the parent cell splits into four daughter cells, each with half the chromosomes of the parent. When two sex cells unite in fertilization, they produce a single cell with a full set of chromosomes, half from the mother and half from the father.

Birth and Beyond

When a baby is ready to be born, the amniotic sac breaks, unloading all that fluid, letting everyone know that birth is near. The hormone **oxytocin** stimulates the uterus to contract, and after (sometimes many) hours of increasing contraction, the opening in the neck of the uterus (the **cervix**) is wide enough to let the baby through. Complications with this process may necessitate a caesarian section, where the baby is removed via an incision in the mother's belly.

The mother's breasts soon start to produce milk—a fatty nutritious substance that is all the baby requires in the way of food for its first months of life.

MAKING BABIES

When the man ejaculates during sexual intercourse, the sperm travels into the woman's uterus. Sperm cells, with their long tadpolelike tails, are reasonably competent swimmers as cells go, but they can't survive too long in the hostile acidic environment they find themselves in. If the timing is right and there is an egg traveling down one of the fallopian tubes, the sperms can detect it by temperature. They surround the egg, and one lucky individual penetrates its outer layers, resulting in the fusion of the chromosomes of both cells and the beginning of a new human life.

By the time it implants, the new cell has already divided many times. Over the next nine months the cell division and specialization continues, and it develops into an embryo, fetus, and eventually a baby. A placenta also develops—an organ that allows the blood of the growing fetus to exchange nutrients and other essential substances with the blood of the mother. An amniotic sac—a membrane containing cushioning amniotic fluid—forms around the fetus.

Respiratory System

Who doesn't enjoy breathing? Our cells need oxygen to release energy. Just as important, we need to get rid of the carbon dioxide that's a waste product of that same cellular process. The air in Earth's atmosphere is mostly nitrogen, which we breathe in and out without anything happening to it, but there's 21 percent oxygen in it, too, which is sufficient for humans (too much oxygen is actually toxic to us).

Nose, Throat, and Lungs

Air enters our lungs via the nose and/or mouth, though the nose generally does a better job. The nose is lined with tiny hairs that trap some of the larger airborne particles that you wouldn't want to inhale, and the mouth doesn't. The airways from nose and mouth join to enter the **trachea,** or windpipe, a tough tube that runs down the throat and into the chest. Here it divides into two main **bronchi** that enter the two lungs.

The lungs are big squashy pink bags. The right lung is shorter and wider and divided into three lobes, while the left has only two; this difference is because of the way the liver and heart fit around the lungs. The bronchi continue to branch off into smaller and smaller tubes (**bronchioles**) inside the lungs, which eventually meet the **alveoli,** the structures where the gas exchange actually happens.

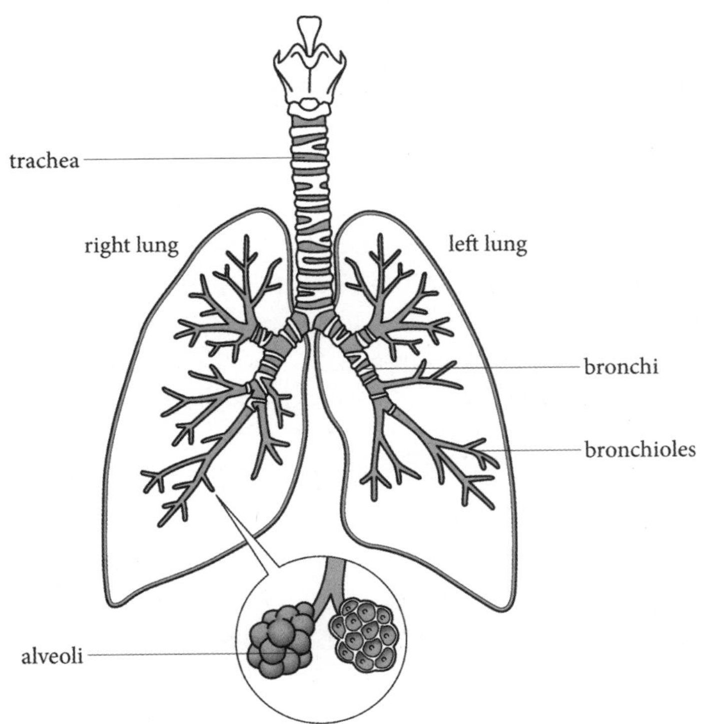

Gas Bags

Alveoli look like little clusters of tightly packed grapes. There are 150 million of them in each lung. Gas exchange takes place on their entire surface area, which amounts to a huge 807 square feet (75 sq m) if you took the trouble to flatten them out. They have a very rich blood supply in the form of numerous tiny capillaries. Here, inhaled oxygen is transferred to the red blood cells, while waste carbon dioxide is released from the blood to the alveoli to be exhaled.

WHOSE AIR IS IT, ANYWAY?

Exhaled air normally has about 4.5 percent less oxygen in it than it did just before you inhaled it, and it contains about 4.5 percent carbon dioxide versus 0.03 percent carbon dioxide in inhaled air. Each breath we take draws in between 4 to 6 quarts (4 to 6 L) of air, and going by the number of humans on Earth and the amount of available air, we can deduce that every breath we take contains at least some molecules that have previously been inhaled by Socrates, Caesar, Beyoncé, or whomever else you care to mention.

Sensory Systems

Our senses are the means by which we get information about our environment. The five "traditional" senses are sight, hearing, smell, taste, and touch, but most biologists agree we have a few more as well, like thermoception (sensing changes in temperature) and equilibrioception (sense of balance). The outside information is detected by receptor cells in the sensory organ (eye, skin, ear, and so on). The receptor cells are all wired into our nervous systems, so the information gathered is translated into impulses along nerves all the way to our brains, where we interpret the data and make decisions about what to do with it.

Some animals have additional senses, such as electroception (the ability to detect electric fields, found in various fish and duck-billed platypuses) and magnetoception, or magnetoreception (the ability to detect magnetic fields, found in migratory birds).

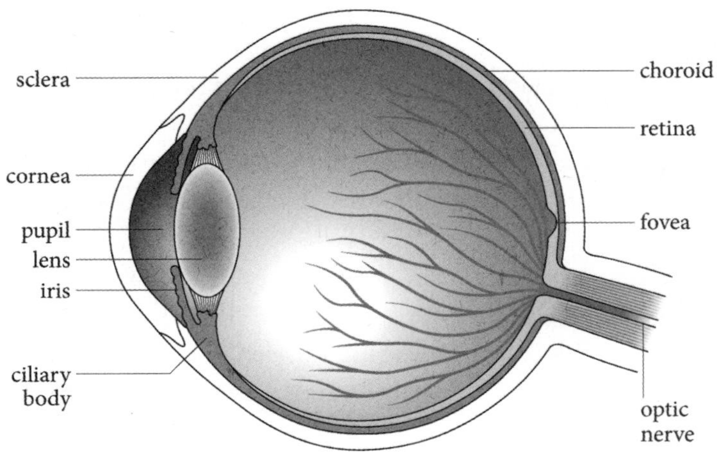

Eyes

The simplest kind of animal eye can distinguish light from darkness. It switches from "off" to "on" when a certain minimal level of light is present, but it has no ability to detect graduations in light levels. Not much use, you might think, but if you're swimming about in the sea and it all suddenly goes dark, that probably means something big and dangerous is approaching and you should leave the area as quickly as possible.

Human eyes are much more sophisticated. We can see a wide range of colors and resolve lots of detail from the picture of visible light that hits our retinas. We can also focus at close range and long distance, can cope quite well in low light, and can quickly detect movement. The human eye consists of several parts.

Sclera. The "white" of the eye is a tough covering that becomes transparent at the front, at a place called the **cornea**.

Iris. The colored part is a round sphincter muscle that contracts or relaxes to change the size of the pupil.

Pupil. The black part is actually just the hole at the center of the iris, through which light passes.

Lens. Once through the pupil, the light radiation hits the lens, a transparent structure that changes shape to focus the light.

Retina. The inside layer of the eyeball that contains the special light-sensitive nerve cells. These cells contain chemicals that react if certain wavelengths of light reach them. They come in three kinds: rods, cones, and photosensitive ganglion cells.

The **rods** are very sensitive to light of all colors, so they help us detect movement quickly—especially at the edges of our vision. **Cones** are sensitive to one of three different colors of light: red, green, or blue. Between them the cones supply our color vision, and they are concentrated at the **fovea,** the part of the retina that aligns with whatever we are looking at directly. Photosensitive **ganglion cells** (only discovered in the 1990s; if you haven't heard of them, that's why) help provide length-of-day information because they detect slow, long-term light changes.

Chemical reactions in the retinal cells act as triggers to adjacent nerves, causing them to "fire," and all these impulses are channeled down the **optic nerve,** which exits out the back of the eye. The point where it leaves gives us an optical blind spot—a point on the retina where there are no light-sensitive cells. We are very good at unconsciously compensating for this, however, so that we hardly ever notice it.

FIND YOUR BLIND SPOT

Here is one of several optical illusions that will prove that you do indeed have a blind spot.

Look at the picture from a distance. Cover your right eye and stare at the right-hand circle (containing the X). Now move your head very slowly toward the image. At some point the left-hand circle (containing the spot) will disappear as its position coincides with your blind spot. You'll see the black-and-white square pattern instead—that's your brain filling in the gaps of what hasn't been seen.

The optic nerves from each eye have a long journey to the occipital lobes of the brain, the parts that deal with sensory information; they're right at the back. Halfway along, the nerve fibers originating from the inner section of each retina cross over and end up in the opposite side of the brain. This crossover means that each brain hemisphere deals with half the information from both eyes.

Ears and Skin

Sound waves enter our ears and cause the **eardrum,** a thin membrane, to vibrate. These vibrations pass through the **ossicles,** which are three tiny bones: the malleus, incus, and stapes (aka the hammer, anvil, and stirrup). They increase the pressure of the sound waves. From there the waves reach the liquid-filled **cochlea** in the inner ear, a coiled-up tube structure that contains the receptor cells that convert physical vibrations into electric signals that travel to the brain for interpretation.

The hearing receptor cells in our ears are hair cells, which have fine projections called **cilia** that detect vibrations in the fluid inside the cochlea. Although different in structure, some of the receptor cells in our skin also detect vibration or pressure. It's often said that snakes and some other animals "hear through their skin."

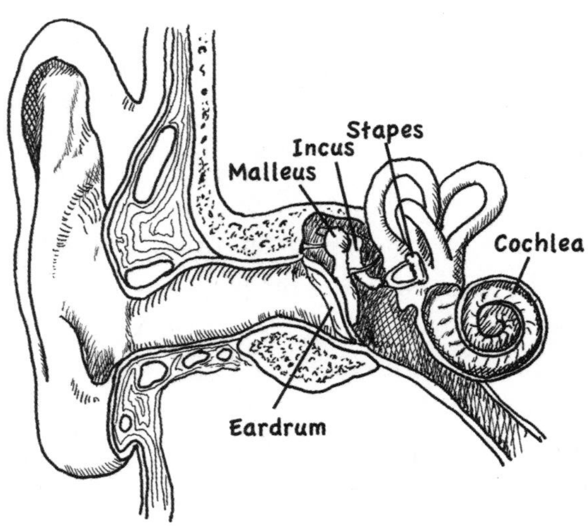

SENSORY MULTITASKING

The inner ears also contain our vestibular system, which helps us maintain our balance. It consists of three fluid-filled looping tubes—semicircular canals. The movement of the fluid inside them as our heads move around tells us which way is up and whether we're about to fall over.

Smell and Taste

The receptor cells in our noses and the taste buds on our tongues are chemoreceptors—they are activated in the presence of certain chemical molecules. For a long time the human tongue was thought to be sensitive to just four tastes—sweet, salty, sour, and bitter—with all the other nuances of flavor actually provided by the nose, which is sensitive to a much wider array of smells (though not nearly so many as some other mammals, such as dogs). A fifth taste sensation was identified in the early twentieth century—**umami,** or "savoriness."

Hormones

Hormones regulate a whole range of bodily processes, many of which happen to be associated with particular feelings as well as with essential processes, like our sleep-wake cycle, growth, reproductive mechanisms, and our physiological response to danger.

Hormones are secreted into the bloodstream, where they travel to other parts of the body and cause cells to alter their metabolism (their chemical processes) in some way. That includes the rate at which they produce hormones—in other words, hormones regulate the production of other hormones.

Types of Hormones

All hormones are molecules, but they vary widely in shape, size, and composition. All are either protein-based (amine or peptide hormones) or derived from fats (steroid hormones). Hormones are released by **endocrine glands,** of which there are several in the body. Some organs are nothing but endocrine glands, such as the adrenal gland, one of which sits on top of each kidney. Other organs, such as the pancreas, count their endocrine functions among a wider array of talents.

Nearly everyone has heard of the hormone **testosterone**. It is secreted by the testes in men and the ovaries in women, topped up by small contributions from the adrenal gland. Men produce about 50 times more of it than women, and it is the key hormone in establishing and maintaining maleness in a human body, from conception to maturity. Embryos all begin with a female-type body plan, but male embryos start to produce testosterone, which turns their genitals masculine. In puberty, testosterone causes changes like armpit hair growth and acne in both sexes, and extra hairiness, voice deepening, and shoulder-broadening in boys. In adults (men and women) it helps regulate and maintain the libido.

Another hormone you've also have heard of is **insulin,** which is secreted by the pancreas. This has a much more specific effect, causing cells in our muscles, fat, and liver to absorb glucose from our bloodstream. Its release is triggered when our blood sugar rises after eating. Without enough insulin the glucose accumulates in the blood, causing a range of symptoms known as diabetes.

How Hormones Work

Different classes of hormones work in different ways. Peptide hormones form a particular folded shape, which fits into a receptor site on the cell membrane of their target cells, thus activating new metabolic processes inside the cell. Steroid hormones actually enter the target cell and bind to receptors in the cytoplasm. The receptors carry the hormone into the cell nucleus, where the hormone activates the genes to make the particular protein that's required to complete the "job" of the hormone.

Cell Biology

You're made of cells. So is your cat, your eyelash mites, and the mushrooms in your omelet—you get the idea. Life forms like us that are made of one or more complex cells make up one of the three primary divisions or domains of life on Earth—the **Eukarya** ("good nucleus"). The other two divisions are both kinds of bacteria, which have a much simpler structure.

Structure of a Cell

In our bodies, cells are specialized to do different jobs. Not all are spherical blobs. Nerve cells are long and thin, red blood cells are like doughnuts with a squashed part rather than a hole in the middle, sperm cells look like tadpoles, and so on. However, most animal cells have the same general working parts, or **organelles,** as shown in the simplified diagram on page 135.

Now, what do all those parts do?

Cell Membrane. This semipermeable skin lets some molecules and chemical ions in and out—but not all, and not always at the same time. Some of this transport is passive and happens by diffusion, but there is also active transport via protein pumps and ion channels.

Cytoplasm. Also called protoplasm, this is made of the organelles of the cell, plus the fluid (cytosol) in which they

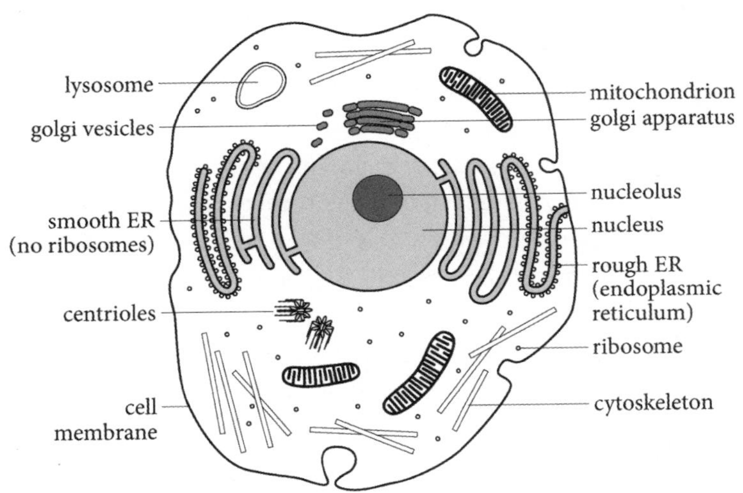

are suspended. The **cytosol** is mostly water with dissolved salts and a few other molecules—it also contains a network of protein fibers, the **cytoskeleton,** which supports the shape of the cell.

Mitochondria. Rod- or spindle-shaped organelles that make the adenosine triphosphate, which supplies the cell's energy needs. They are membrane-bound and contain a huge number of different types of protein. The theory of endosymbiosis says that mitochondria were originally free-living bacteria that later became engulfed by other cells.

Ribosomes. Small round organelles that manufacture proteins from amino acids. They contain ribosomal RNA, which provides the code needed to put the amino acids together in the right order to make specific proteins.

Endoplasmic reticulum. A network of membrane-bound tubes and vessels inside which various functions take place, including synthesizing fats, neutralizing toxins, and "receiving" newly formed proteins from ribosomes.

Nucleus. The largest organelle in the cell. It is bound in a nuclear membrane through which some small molecules can pass. The cell's DNA lives within the nucleus and is duplicated there when the cell divides. The nucleolus is where ribosomal RNA is made, which is the main constituent of ribosomes.

Golgi apparatus. Collects newly synthesized molecules from the organelles that make them and secretes them out of the cell.

Centrioles. These barrel-shaped organelles have an important role in cell division.

Lysosomes. Contain enzymes to digest the remains of "dead" cell components and any virus particles, bacteria, or other relatively large chunks of organic matter that the cell has engulfed.

The Plant Cell

Cells in plants have a lot in common with animal cells, but there are, of course, many differences, too, as the diagram on page 137 illustrates. One of the most obvious differences is the presence of a tough cell wall that surrounds the cell membrane and often gives the cell a more rigid, angular shape. Another is the presence of **chloroplasts,** small green organelles that carry out the process of photosynthesis. A third difference is that plant cells often contain very large

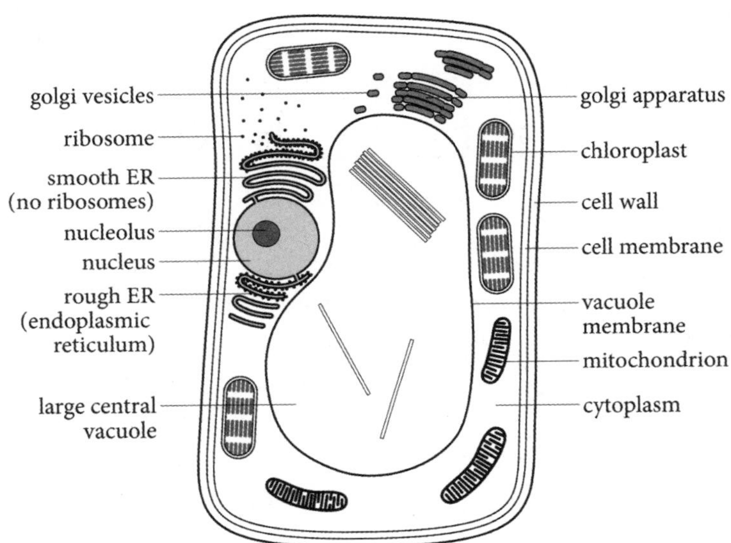

vacuoles—membrane-bound water-filled compartments that deal with waste products and help keep the cell shape rigid as well. Animal cells may contain small vacuoles, but in plants they can take up most of the space inside the cell.

Photosynthesis

Solar panels? Plants invented them first, or at least came up with a way to harness the light energy of the sun to create their own energy: **photosynthesis.** The light of the sun is used to power a chemical reaction between water and carbon dioxide, which makes a useable energy-releasing food for the plant and releases oxygen as a by-product. Blue-green algae, or cyanobacteria, are the simplest organisms that carry out this process. They were among the first life forms to appear on Earth, with signs of their existence dating back 2.8 billion years. They were probably responsible for establishing the oxygen-rich atmosphere we know and love today.

As we saw in the previous section, modern plants contain organelles called chloroplasts in their cells. They look suspiciously like cyanobacteria and support the theory of endosymbiosis—that other cells engulfed the cyanobacteria, where they continued to photosynthesize and to supply food that the engulfing cell was able to use, thus forming the first complex eukaryote cells.

Today modern plants, algae, and cyanobacteria collectively trap a jaw-dropping 100 terawatts of energy from the sun (or 100 terajoules per second), which is more than six times as much as is consumed by the entirety of human civilization.

They also use up a hundred billion metric tons of carbon from atmospheric carbon dioxide. Without them, we would be comprehensively doomed.

How Photosynthesis Works

The chemical equation for photosynthesis looks like this:

$$6CO_2 + 6H_2O \text{ (+ energy)} \longrightarrow C_6H_{12}O_6 + 6O_2$$

The energy comes in the form of photons. Light energy from the sun is captured by the chlorophyll pigment inside the plant's chloroplasts. The carbohydrate that's produced is glucose, the simplest form of sugar, and the plant stores that in its tissues, while the oxygen is released back into the atmosphere.

ADP, ATP, and Energy

The photosynthesis reaction uses up energy and produces a more complex compound from the reactants. Reactions of this kind are called **anabolic.** The reverse of this, where a compound is broken down into simpler products and energy is released, is called **catabolic.** The catabolic reaction of aerobic respiration, which happens in plants and animals alike, is the way living cells use oxygen and glucose to release energy to use as required. The reaction as an equation looks like this:

$$C_6H_{12}O_6 + 6\,O_2 \longrightarrow 6CO_2 + 6H_2O \text{ (+ energy)}$$

If you've been paying attention, this should look familiar. It's the reverse of the photosynthesis reaction. The key difference is the nature of the energy. Energy input in photosynthesis is light energy, the energy output from respiration is chemical, and it is carried by the molecule ATP (adenosine triphosphate).

ATP consists of the nucleotide adenosine (which also happens to be a component of DNA) with three phosphate groups (monomers of oxygen and phosphorus) attached. When respiration occurs, molecules of the stable ADP (yep, adenosine diphosphate) which have two phosphate groups, are turned into the unstable ATP by the addition of a third phosphate group, via an unstable high-energy chemical bond. When energy is required, it is released by the breaking of that bond, which causes the breakdown of ATP back into ADP. See, it is all about the chemistry.

GIVE AND TAKE (AND TAKE SOME MORE)

We often imagine an oversimplified picture of gas exchange in the natural world, with plants using up carbon dioxide and releasing oxygen, and animals using the oxygen and releasing carbon dioxide, and on and on forever in happy equilibrium. However, plants use up oxygen, too, in the same respiration reaction as animals. So the main difference between animals and plants is actually that plants synthesize their own food supply, which happens to use up atmospheric carbon dioxide.

Genetics

Genetics is the science of genes, heredity, and variation in living organisms.

Chromosomes

The DNA that lives in the nucleus of each cell in your body (apart from the ones that don't have a nuclei, like your red blood cells) is one long molecule mixed up with proteins and is called **chromatin.** When a cell is about to divide, the chromatin is organized into strands called **chromosomes.** When they are about to divide, they look like tall thin X's, with a join in the middle where they separate. Humans have 46 chromosomes in 23 pairs. The pairs all match, except for one—in men only. In women both of the "sex chromosomes" are X chromosomes, but in men one sex chromosome is an X and the other is the much smaller Y. A full set of chromosomes is called a **karyotype.**

Each chromosome is made of numerous genes. A **gene** is a length of DNA that holds the code for making a specific protein, as well as coding for whether it is "switched on" or "switched off."

The chromosomes contain all the instructions you need to make all the proteins that form your adult body. From the

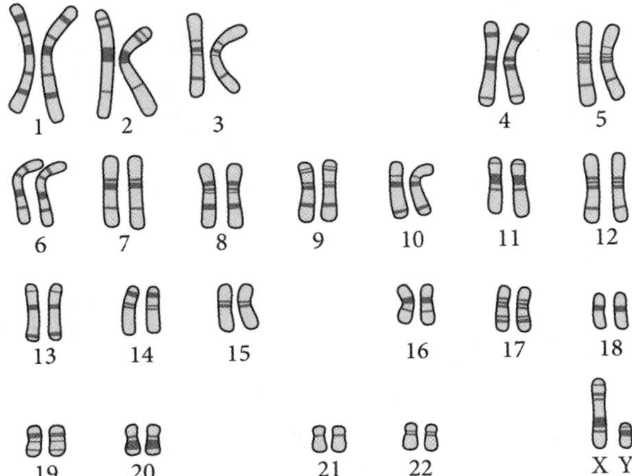

moment of fertilization, the chromosomes in the single cell, which results from the fusion of egg and sperm, are at work coding new proteins that make new cells and carry the instructions for what the cells need to do. Because half your chromosomes come from your mother and half from your father, when put together they make a brand new combination of genes.

When Chromosomes Divide

As we saw in the Reproductive System section (page 120), there are two kinds of cell division. There's mitosis, when a cell simply duplicates itself and all its chromosomes, and meiosis, which happens with sperm and egg cells, producing from each parent cell four daughter cells with half the chromosomes of their parent cells. These cells are called haploid, and each contains 23 chromosomes, one of each pair.

During meiosis the cells divide twice, so one cell becomes four. On the first division the two new cells still have 46 chromosomes in 23 pairs, but the chromosomes themselves are rearranged—a process called **recombination.** So parts of the first member of chromosome pair 1 get combined with bits of the second member of pair 1, and so on, creating two cells with a different set of genes from their parent cell. Then the second division occurs, resulting in four haploid cells. The recombination means that no two haploid cells ever contain the same combination of genes.

Inheritance

Most of us can see some of our parents' features in our own faces. Most of our features are shaped by the influence of many genes (we have something like 20,000 genes altogether). To understand how inheritance works, it's easiest to look at a trait controlled by a single gene—the trait is either present or absent depending on which versions—or **alleles**—of a particular gene you have. Traits like this are called **Mendelian,** after their discoverer Gregor Mendel. Common Mendelian traits in humans include cleft chin, facial freckles (most obvious in pale-skinned people, but anyone can have them), the ability to flex your thumbs backward, and whether you have free or attached earlobes (i.e., dangly earlobes or earlobes joined straight to the side of the head).

PHENOTYPE AND GENOTYPE

Time to explain these two tricky terms. Your **phenotype** is your outward appearance. In the case of freckles, your phenotype would be "freckly" or "not-freckly." Your genotype describes which alleles (see page 146) you have with regard to that particular trait—whether you have two f alleles, two F alleles, or one of each. If both your copies of an allele for a trait are the same, your genotype is **homozygous** for that trait—if they are different, you have a **heterozygous** genotype for that trait (homo = the same; hetero = not the same). As you'll see, it's not always possible to tell a genotype by looking at the phenotype.

Alleles

A Mendelian trait is either **dominant** or **recessive**. If it's dominant, the trait will be expressed whether you have one or two copies of the allele. If it's recessive, you need two copies. Let's take the freckles example. This gene resides on chromosome 4, so on each of your copies of chromosome 4, you have either the dominant having-freckles allele or the recessive no-freckles allele. Let's call them F and f, respectively.

Bob and Mary both have freckles; Jim and Susan both lack them. From that we can deduce that both Jim and Susan have the genotype ff—two copies of the no-freckles allele.

Jim = ff

Susan = ff

With Bob and Mary we can't be so confident. They have the same phenotype—freckly, but because F is dominant over f, they could have different genotypes—either one copy of F and one of f, or two of F.

Bob = FF or Ff

Mary = FF or Ff

If Jim and Susan have babies, they will all have the same non-freckly phenotype and the same ff genotype as their parents—because neither Jim nor Susan have any F alleles to contribute. If Bob and Mary both happen to be FF, they too will have children with the same phenotype (freckly) and genotype (FF) as themselves.

ff parent x ff parent = 100 percent ff offspring

FF parent x FF parent = 100 percent FF offspring

What if Bob is FF and Mary is Ff, though? Because each of their children will inherit an F allele from Bob, they will all have the freckly phenotype. But because half of them will get Mary's F allele and the other half her f allele, they will be half FF and half Ff.

FF parent x Ff parent = 50 percent FF, 50 percent Ff offspring

The most interesting outcome is if both Bob and Mary are Ff. There are four possible combinations from this—FF, Ff, Ff the other way around, and ff.

Ff parent x Ff parent = 25 percent FF, 50 percent Ff, and 25 percent ff offspring

To explore the final possible combinations, we need our couples to do a bit of clandestine wife swapping. Let's say we've established that Bob is FF and Mary is Ff. Bob and Susan's love children will each inherit one of Bob's two F alleles and one of Susan's two f alleles, so they will all have a freckly phenotype and an Ff genotype.

FF parent x ff parent = 100 percent Ff offspring

Half of Jim and Mary's children will inherit Mary's F allele, and the other half will inherit her f allele, while all will get one of Jim's two f alleles.

Ff parent x ff parent = 50 percent Ff and 50 percent ff offspring

From this we can deduce that (assuming both illicit pairings result in a child and everyone concerned knows about Mendelian inheritance) Bob's and Susan's affair has a 100 percent chance of being detected, while Jim's and Mary's indiscretions have only a 50 percent chance of discovery.

Inherited Diseases

Not all Mendelian traits are innocuous things like freckles. Some devastating diseases are passed on this way, too. One is Huntington's chorea, a slowly progressive and fatal disease of the nervous system that develops in early middle age. It is a dominant trait, meaning that anyone who has a copy of the allele will develop the disease, and anyone with one affected parent has a 50 percent chance of developing it. Genetic testing means that problems like this can be detected and affected people can make an informed decision whether to have a family. On the other hand, cystic fibrosis is a recessive trait; you can inherit it from two unaffected parents if they are both heterozygous for it.

Alleles on the X chromosome work in a slightly different way. The X and Y chromosomes are not a matching pair—the Y is much smaller, with far fewer genes on it. So in men (who have one X and one Y) an allele for a recessive trait on the X chromosome will always be expressed in the phenotype, while in women (with two X's) it may not. An example is the blood-clotting disorder hemophilia. If a man inherits an allele for this from either parent, he will develop hemophilia, but a woman will only develop it if she inherits the allele from both parents. (Genetic disorders are not the same as congenital conditions. The latter are not inherited, but are caused by something going wrong with the chromosomes prior to or just after conception. Down's syndrome is an example of a congenital condition.)

Reproduction and Cloning

As we've seen, in sexual reproduction genetic variety is introduced, both during meiosis, when each parent's chromosomes get chopped up and mixed up (recombination), and at fertilization when half of the chromosomes from each parent are combined to form a brand-new full set.

The alternative to sexual reproduction is cloning, where a single parent produces an exact copy of itself, with the same chromosomes. What little variety there is between parent and offspring can only be introduced in the form of genetic mutations that may or may not occur.

The cells in our body reproduce by cloning, and single-cell organisms do it, too. Some more complex life forms also retain the ability to naturally clone themselves, while retaining the option of sexual reproduction as well. In animals this process is called **parthenogenesis** and is easily observed among various kinds of insects. Problems with greenfly or blackfly in your garden? Those little aphids carpeting the rose stems are all clones, born alive and churned out by the females all through the summer, with no need for any male involvement at all.

Cloning in Mammals

Scientists have been working on cloning mammals of various species for decades. The first viable mammal clone was Dolly the sheep, born in 1996. Her creation began with the removal of an unfertilized egg from the ovary a female sheep—Sheep A. The nucleus was removed and replaced with that from a cell from the mammary gland of a different female sheep—Sheep B. This new cell was placed in the uterus of yet another female

sheep—Sheep C, I suppose, where it grew and developed. In due course Dolly was born, a clone of Sheep B. Since Dolly, scientists have successfully cloned various other mammals, including cats, cows, horses, and rhesus monkeys, though the failure rate remains very high.

WHY CLONE?

Cloning has a strong whiff of "Frankenstein science" about it, second only to genetic engineering (directly manipulating an organism's genetic makeup) in its ability to inspire unease. It's hard to imagine a scenario outside of science fiction where it would find widespread acceptance. However, cloning does have its uses—real and potential. Cloned plants have been used in agriculture for many years. We could clone new organs and solve the problems of insufficient registered donors and incompatible hosts. Endangered species could be saved from extinction. We could even make a theme park full of dinosaurs. . . . What could possibly go wrong?

Evolution

Until we invent a reliable forward-and-backward time machine, we can't be sure how life first began on Earth—the same problem that we have when contemplating the beginning of the universe (if it had one). We can theorize scenarios and attempt to re-create them in the lab, but we may never have the same confidence about our theories as we do about what is observable in the here and now. For that reason, many people prefer a supernatural explanation for problems like this, but this is a science book, so we're going to stick with looking at naturalistic ideas.

Abiogenesis

Based on what's emitted from volcanoes today, we can be reasonably confident that the prelife atmosphere of Earth would have contained a mixture of gases, mainly water vapor but also

including methane, ammonia, and hydrogen. Apply energy to this mixture with a lightning strike and reactions could occur between the gases, producing simple organic molecules.

The scientists Stanley Miller and Harold Urey tested this idea by applying electric sparks to a mixture of the aforementioned gases. The result was the generation of quite a few different organic molecules, including sugars, fats, and amino acids. All the building blocks necessary to create nucleotides—the basis of the replicating molecules DNA and RNA—were formed. Other subsequent experiments have yielded similar results, and some have produced whole nucleotides.

RNA World

Our cells contain RNA (ribonucleic acid), which has various functions, including acting as a messenger to carry the coded information in DNA from the cell's nucleus to the ribosomes, which make proteins. Like DNA, it is a chain of nucleotide molecules—four different kinds arranged in various different orders, so it has similar information-storage properties and

self-replication abilities to DNA. However, it has only one strand rather than two, so is a smaller molecule, able to go places where DNA cannot.

The theory of RNA World, first proposed by Walter Gilbert in 1986, hypothesizes that the first "life" on Earth was a large quantity of free-living RNA molecules of various different configurations, all replicating away. RNA can behave as an enzyme, potentially helping to catalyze (speed up) other chemical reactions, which could provide it with additional chemical building blocks. Those configurations of RNA that were most successful at replicating themselves and at breaking down other chemicals would survive better than the rest.

The theory says that in due course the double helix of DNA would have formed and survived well because it is inherently much more stable than RNA. From free-living RNA and DNA to the first simple cells is not so much of a stretch—a virus is little more than some DNA or RNA strands held inside a protein coat, and a bacterium a little more complex again. The theory of **endosymbiosis** (when one simple cell engulfs another and the engulfed cell continues to "work" for both cells) is the popular explanation for the origin of complex cells.

The Evolution of the Eukaryotic Cell

We already touched on how the first complex eukaryotic cells—the kind that has a nucleus and other organelles, and the kind that you and I are made of—evolved. Time for a closer look at the theory of endosymbiosis. We can see many endosymbionts around today—organisms that can only live inside another organism, with both benefitting from the

DOMAINS OF LIFE

Before we go any further, we need to take a quick look at the primary divisions of life on Earth—or at least cellular life (in other words, viruses and other simple "possible-life" entities are excluded).

Cellular life is divided into three domains—Archaea, Bacteria, and Eukarya. Eukarya are the ones made of complex cells—at least one cell but often many, each cell containing a nucleus and lots of other organelles. Archaea and bacteria are simple single-celled organisms, smaller in size and with no nucleus and very few organelles. Modern bacteria, including cyanobacteria, belong in the domain Bacteria. However, one group of somewhat oddball bacteria were found to be profoundly different, and they are now classified in an entirely separate domain—the Archaea (a single individual within the domain is an archaeon).

association. Examples include the bacteria that live inside the roots of pea plants, obtaining nitrogen for their hosts, and friendly bacteria that live inside the intestines of animals, helping them metabolize their food.

Endosymbiosis

The theory of endosymbiosis says that an archaeon engulfed some bacteria cells in the first stage of evolution of the eukaryotic cell—the archeon became a eukaryotic cell and the bacteria cells became mitochondria or choloroplasts inside it. This idea is borne out by the fact that DNA in our cells' nucleus is similar to that of Archaea.

So the mitochondria in a complex plant or animal cell are thought to have descended from free-living bacteria. Likewise, chloroplasts in plant cells probably originated as free-living cyanobacteria—aka blue-green algae. Both chloroplasts and mitochondria contain their own DNA, which is different from that found in the nucleus of the cell they're in. Their membranes are structured like those of bacteria. Both contain their own ribosomes, which are bacteria-like and distinct from the ribosomes at large in the cell itself. They reproduce themselves by dividing, just as bacteria do. It has been suggested that other cell components were also once free-living organisms, but there is not so much evidence for them as there is for mitochondria and chloroplasts.

And the Rest?

There's a lot more to a eukaryotic cell than its mitochondria and (in plants) its chloroplasts. The nucleus, the most important organelle of all, is of uncertain origin, but some biologists believe that the DNA it contains is that which belonged to the "engulfer" archaeon cell when endosymbiosis occurred, while others think it came from a separate virus that also became engulfed.

Ribosomes are the only organelles that are found inside Archaea and Bacteria cells. Their origin therefore predates that of the eukaryotic cell itself, and indeed the free ribosomes in a eukaryotic cell's cytoplasm are like those of an archaeon, while the ribosomes inside mitochondria and chloroplasts are like those of a bacterium.

Other cell organelles are generally similar to each other and are thought to have started out as inlets of the main cell membrane, gradually becoming more specialized over time. The driving force behind their changes was simply how well they helped the cell to survive and propagate itself—natural selection, which we'll look at in more detail in the next section.

Mutation and Natural Selection

From such simple beginnings came the whole diversity of life on Earth, guided only by the natural processes of genetic mutation and natural selection—in other words, evolution. Like abiogenesis, the theory of evolution is much challenged by those who prefer a supernatural explanation for how we got here, but evolution has much more solid science to support it than abiogenesis does. To argue against it is akin to disputing the theory of gravity, in terms of the amount of evidence that you would need to explain it away by some other means.

How It Works

Evolution happens when there are two conditions. The first is that the population of organisms has variety—all individuals are slightly different. The second is that there are environmental pressures on the animals—only those that function best in their environment will survive. For a real-world example, consider a population of gazelles. Lions eat the gazelles, so the gazelles that run faster are more likely to escape the lions and live long enough to breed, passing on their fast-running genes to their offspring. So the proportion of speedy runners increases in the population as a whole.

Variety in a population is introduced by mutation. When a single-celled organism reproduces, it splits in two and its DNA is replicated, but small mistakes—mutations—may occur in that replication. If that happens, the new cell is genetically different from the parent cell. In species that reproduce sexually, with each organism's genes coming from two parents, more genetic variety is added. The complete set of genes, or **genome,** are the instructions for "building" the organism—a new genome produces an organism that's built differently from its parent or parents.

The word "mutation" has negative connotations because most of the mutations we know about are harmful. Mutation is random and undirected—a mutation is as likely, if not more so, to have a damaging or neutral effect as it is to having an advantageous one. That's where the other part of the process—natural selection—comes in. This process ruthlessly weeds out all those individuals that aren't well equipped for survival, so that only advantageous mutations are preserved in a population—the survival of the fittest. We do a similar thing with our domestic animals and cultivated plants, breeding only from those that show the traits we want to keep and develop. This process—artificial selection—is responsible for the huge variety of dogs, chickens, and chrysanthemums that live on Earth today.

Divide and Conquer

Where one species occupies a large, environmentally diverse area, there can be different selective pressures on the different parts of the population. Given consistently varied selective pressures over a sufficiently long time, a population can split

THE PEPPERED MOTH

One of the most famous examples of natural selection in action concerns a fairly large fluffy moth called the peppered moth. Most peppered moths are a pale gray, but a genetic variant with blackish wings occurs from time to time. In industrialized areas in the late 1800s and early 1900s, the black form was found to be much more common than the pale form.

Peppered moths fly at night, and in the daytime they rest on tree branches. The dappled gray color of the pale form provides good camouflage against a lichen-covered branch, while the black form stands out like a red wine stain on a sheepskin rug and is quickly spotted by predators. However, when industrial soot coated city trees, the tables were turned, and it was the black form that had the good camouflage. We've cleaned up our act quite well since then, and the black form is rare again.

---

into two distinct types. One of the most widespread species of birds in the world is the peregrine falcon. Peregrines live on all continents except Antarctica and in all kinds of environments except the poles, the highest mountains, and the deepest rain forests. However, not all of them are the same. Peregrines from the Arctic are larger and paler than those in the tropics. Being pale is good in snowy environments, and being large is better than being small when you need to retain heat in cold climates; therefore, the larger and paler birds survived best in the North and came to dominate the peregrine populations there.

The next stage is when a population becomes physically divided, with no interchange between the two. The mosquito *Culex pipiens* is common in the UK. At some point in the last century, some of these mosquitoes found their way into the London Underground system, where they survived very well, feeding on the blood of commuters. Recent study of these "underground mosquitoes" has found that they show very different behavioral patterns from their overland counterparts and will not willingly interbreed with them. The "underground mosquito" is now classified as a different species—*Culex molestans.*

An Inexact Science

We humans love to categorize things. When it comes to the world of animals, the species is the preferred unit of organization. We define them by whether or not they can breed with each other—horses and donkeys are clearly similar and will

LIVING FOSSILS

Environmental change means species must adapt or die. However, sometimes there is little or no change in a certain environment for a huge amount of time, so the pressure on the organism is to stay just as it is. The best-known example is a big, ungainly fish called the coelacanth. Modern coelacanths haven't changed much from their fossil relatives from before the time of the dinosaurs, more than 400 million years ago.

---

mate, but their offspring is infertile, so we can safely consider them different species. Go back in time far enough and you will find an animal that is an ancestor to both horses and donkeys, but there is no exact moment when the two lineages split. The yellow wagtails in Britain look very different from the ones in Greece, but they will still successfully interbreed, so they are not separate species—yet. Give it a few more eons of separation and they could well be. At the moment they are classed as separate subspecies.

The division of a species into two species is called a **speciation event.** Because huge timescales are involved, it's very rare for us to actually observe one, though we stand a better chance among animals that have lots of generations in a short span of time. Scientists have seen the appearance of new species of bacteria in the lab, including one with the handy ability to digest nylon. And by studying differences in DNA, we can determine that the most recent ancestor that we share with our closest living relative—the chimpanzee—lived about 5 to 7 million years ago.

Population

There are currently well over 6 billion members of the species *Homo sapiens* walking around on planet Earth. Haven't we done well? In sharp contrast, the total wild population of Spix's macaw, a pretty blue parrot from Brazil, was—until recently—one. (It is now none; there are a few dozen in captivity still.) The vast majority of species that have ever existed are now extinct. While the process of evolution is the means by which species adapt to changes in their environments, it takes time, and when environments change very quickly, most species don't adapt in time, especially when their population is small.

What Regulates Populations?

Populations of living things are constrained by the resources available in their environment. A "resource" could mean something as simple as physical space (think of barnacles on a rock, or trees in a dense forest). It also covers food, water, suitable breeding sites, places to hide from predators, and anything else the species in question needs to survive. The environment will have enough resources to sustain a certain population size of each species—this is its **carrying capacity.** The population won't grow beyond the environment's carrying capacity, because some individuals will die or fail to breed due to lack of resources.

What Changes Populations?

Environmental changes will affect the carrying capacity of that environment. A drought strikes and there's less water to go around, so animals die of thirst, and the population stabilizes at a lower level. It may work the other way around, too. As the Agricultural Revolution spread west across Europe, the newly created farmland habitats allowed the populations of birds, like the collared dove, which favors open habitats, to increase and spread. Of course, in a complex and dynamic system like planet Earth, things can get complicated and consequences can be difficult to predict. For example, a forest fire may kill some of the animals that live there but may also create a new and better habitat for some of the survivors, so numbers could bounce back quickly.

Bottlenecks

In species that reproduce sexually, reducing the population too much causes **inbreeding depression,** when all of the individuals able to breed are very closely related. This is bad news, because inbred organisms are more likely to have health problems—the lack of genetic variety means that a disease affecting one individual is likely to also affect the others. The population of the northern elephant seal fell to just 30 individuals in the 1890s—though numbers have since recovered, all of the thousands of northern elephant seals alive today are very closely related to one another.

Another species to have gone through a population bottleneck like this is us humans, the *Homo sapiens.* Some 70,000 years ago a supervolcano in Indonesia erupted, causing a mini ice age and killing off a huge proportion of the humans on Earth at the time, leaving no more than 10,000 potential breeding couples (possibly far fewer). Today the 6 billion humans on Earth exhibit a lot less genetic variability than the 150,000 or so chimps and 125,000 gorillas.

Predation

All animals consume other organic material. Many eat just plants—the **herbivores**—and some are **omnivores,** which eat both plants and animals. Then there are quite a lot of them that only eat other animals, which they catch and kill. These are the **carnivores** or **predators.** When we think of a predator, we probably imagine something big, scary, and well-equipped with pointy claws and teeth. However, a robin is as much a predator of worms and spiders as a leopard is a predator of antelopes.

The Food Web

You probably remember the concept of a food chain. Here's an example:

Grass > Rabbit > Fox

The rabbit eats the grass; the fox eats the rabbit. Let's flesh it out (so to speak) a bit more.

We now have a web rather than a chain, showing how energy and nutrients cycle from organism to organism. The grass gets its energy and nutrition from the sunlight and the Earth. The rabbit eats the grass; the fox eats the rabbit. The fox has no predators, but when it dies, its body is broken

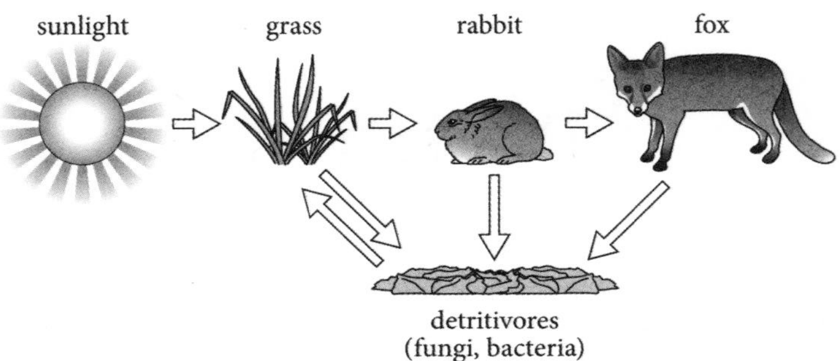

down by decaying organisms, or **detritivores,** like fungi and bacteria. The same thing happens to grass and rabbits that die rather than get eaten. The detritivores release nutrients in a form that the grass can take up. You can add as many other organisms as you like to a food web—this one, set in a typical countryside, could easily include oak trees, stinging nettles, assorted insects, flycatchers, hawks, wood mice, and badgers.

Predator-Prey Relations

It's fairly obvious that with the simple food web shown here, if you take away the rabbits, you can't have any foxes. For that reason, the foxes can't wipe out the rabbits. Because each fox needs lots of rabbits, the foxes will die out before they have eaten every last rabbit. In real life, of course, there aren't just rabbits, there are mice and birds and worms and many other things that the foxes can eat instead, but it is still very unlikely that a predator will ever kill off every last individual of any of its prey species. If rabbits became rare, the foxes that survived best would be the ones that were good at catching prey other than rabbits.

In the average ecosystem, predators and prey have evolved together over millenia, and their populations fluctuate in a predictable pattern. In the Arctic, snowy owls mostly eat lemmings. In years of high lemming population numbers, most of the baby owls will get enough to eat and the owl population will grow. If lemming numbers crash, a lot of baby owls will starve. Over time their numbers fluctuate like this:

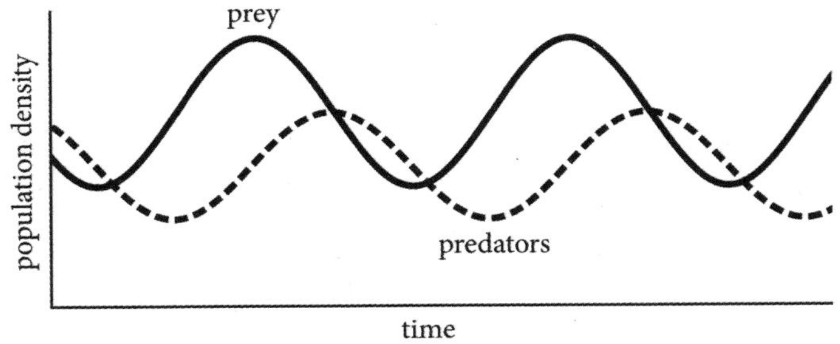

If we took away the owls, the lemmings' line of the graph may go up higher, if their numbers have not already reached the carrying capacity of the habitat. However, in natural situations, predators' numbers are much more strongly controlled by prey numbers than the other way around.

Predators that are moved from their native lands to new places can have much more damaging effects. In a bid to control the rampant population of introduced rabbits in Australia, people introduced stoats (a type of weasel) to eat them. A sensible enough move, you might think—stoats are expert rabbit hunters in Europe, after all. However, the introduced stoats were soon happily decimating the small native animals, none of which had ever coexisted with stoats, so unlike the rabbits, they had not evolved any kind of defense against them.

Extinction

We have already seen that most of the species that ever lived are already extinct. Life has existed on Earth for about 3.7 billion years, and over that time things have changed here a great deal. We have had ice ages, supervolcano eruptions, asteroid strikes, and plenty of tectonic activity changing the shapes of our continents and causing earthquakes and other mayhem in the process. All of these events have placed massive pressure on Earth's living passengers and have put an end to the existence of more than 99 percent of all species, ever.

Defining Extinction

Extinction is one of the simpler concepts to get your head around in biology—when every last member of a species has died, that species is extinct. Species that are not extinct are called **extant.** If there are some individuals of a species left but they are not able to breed (for example, they are all males), that species is functionally extinct. Several species are extinct in the wild but have living representatives in captivity—Spix's macaw, which we mentioned earlier, is an example.

Causes of Extinction

There is an urban legend that a small flightless bird called the Stephens Island wren was completely exterminated within a few weeks by a single cat brought to the island by sailors in 1894. While the veracity of some details of the story are disputed, there's no doubt that this particular environmental change was too fast for the small wren population to adapt before all were gone. Such is the cause of all extinctions—a change too fast and too severe for the species to adapt to, whether it be the arrival of a new predator, a climate change, a volcano eruption, the destruction of a large swathe of habitat to build a new hotel, or any number of other disasters. If a large part of a population is destroyed, the remaining populations may be too fragmented to connect up again and too small to survive alone.

Mass Extinctions

The fossil record shows us that since life began on Earth, there have been five periods of time when extinction rates went up way beyond the usual level. These mass extinctions include the one 165 million years ago that killed all of the dinosaurs, and an even more devastating event 250 million years ago that wiped out 96 percent of all marine species. Some biologists say that the current high rate of extinction due to human activity warrants inclusion as the sixth mass extinction.

Index

Page numbers in **boldface** indicate diagrams and illustrations

Gravitational collapse • Expanding uni

ang • Laws of motion • Transverse wa

lectromagnetic spectrum • Planck's c

Alpha particles Geothermal power •

nergy • Conduction Convection • T

ble • Chemical elements • Transiti

arth metals • Alkali metals • Atomic

Electronegativity • Planck's Oxida

Particulate matter • Galvanization •

NA • Nucleobase • Monomers •

ymphocytes • Erythrocytes • Plasma

Interstitial fluid • Femur • Patella •

mooth muscle • Epidermis Sebaceou

Dendrites • Cerebral cortex • Somat

nzymes • Mitosis • Amniotic sac Alv

ornea • Ganglion cells • Rods and co

Ribosomes • Endoplasmic reticulum

aryotype • Phenotype • Genotype •